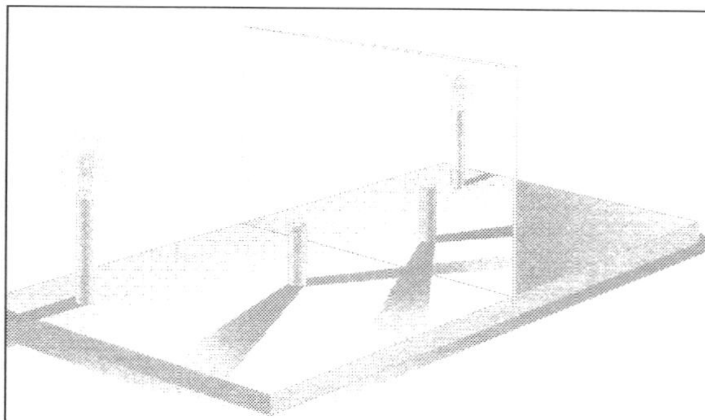
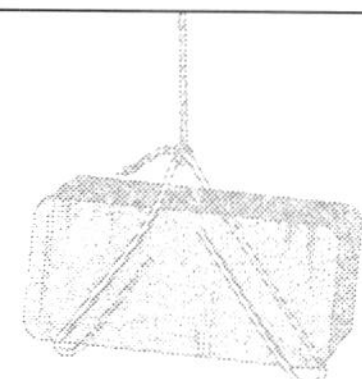
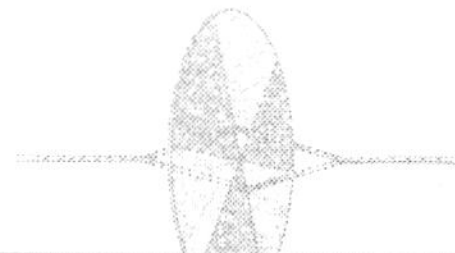

PHYSICS

the Waldorf Way

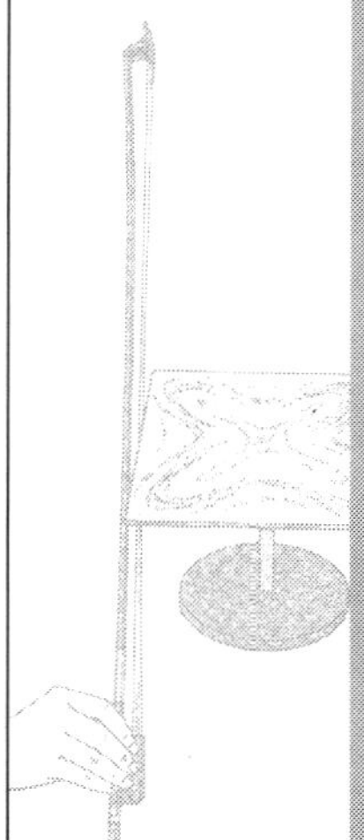

Grade 6

A Manual for Teachers

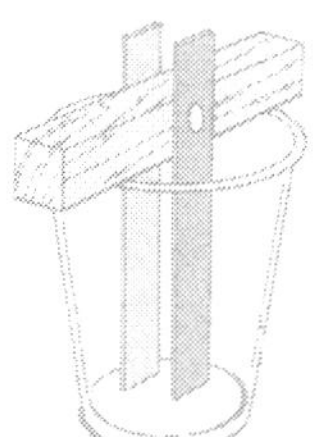
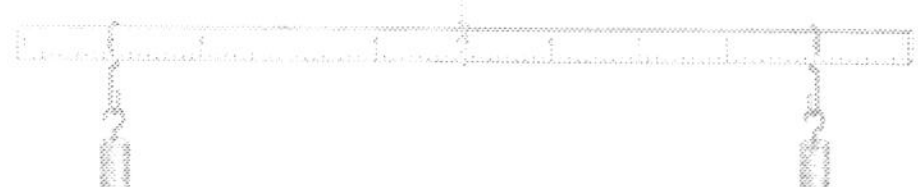
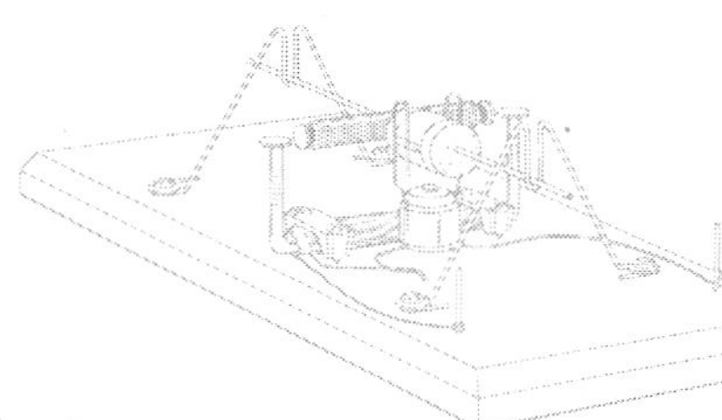

Roberto Trostli

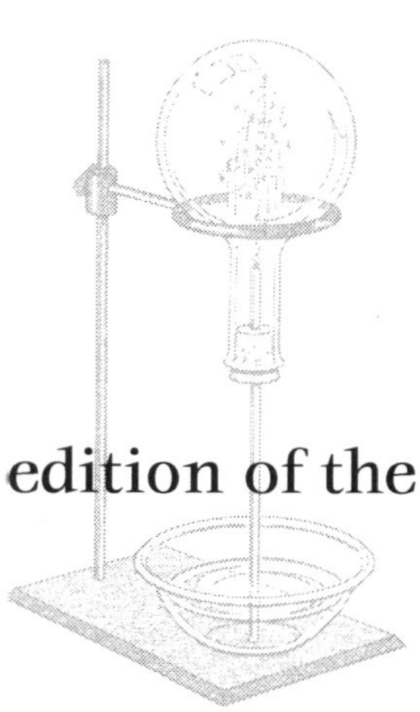
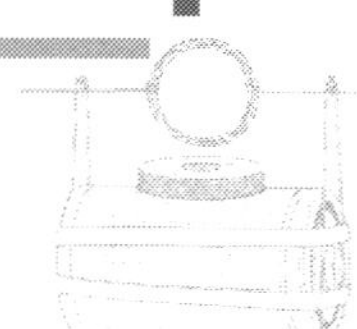
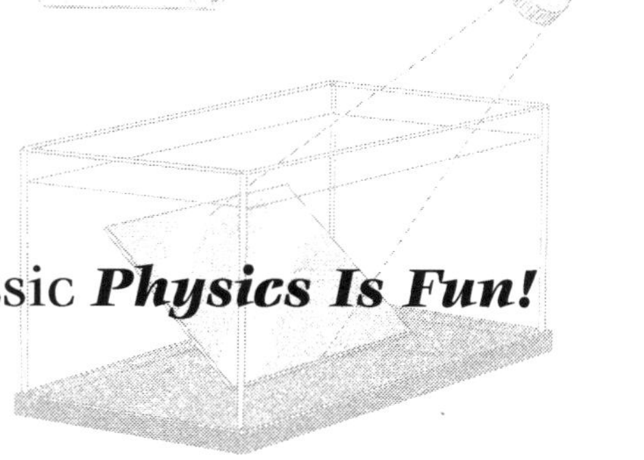

A newly revised edition of the classic ***Physics Is Fun!***

LYSTRA BOOKS
& Literary Services

Physics the Waldorf Way, Grade 6: A Manual for Teachers
A Newly Revised Edition of the Classic *Physics Is Fun!*

Orders can be placed through the website www.waldorfphysics.com.

ISBN: 978-0-9861516-2-0
Library of Congress Control Number: 2015941437
Education: Science

The original edition, *Physics Is Fun! A Sourcebook for Teachers* by Roberto Trostli, was published in 1996 by Octavo Editions through the generosity of Thomasina Webb.

Illustrations by Louis Pappas

Book design by Kelly Prelipp Lojk

LYSTRA BOOKS
& Literary Services

Editorial work and publication management by Lystra Books & Literary Services, LLC, 391 Lystra Estates Drive, Chapel Hill NC, 27517, lystrabooks@gmail.com.

Printed in the USA

This book is dedicated to
Donna Marie Trostli, who has been
my steadfast companion, inspiration,
and support for over three decades,
and to Garrett Anthony Trostli, who
seems to have inherited his grandpa's
love of learning.

Preface

Many years ago, when I was a new and inexperienced Waldorf teacher, I was particularly anxious about teaching physics to my sixth graders. My own science background was sketchy at best, and there seemed to be few resources to consult. I looked around for experiments and activities, but I had no idea what would be appropriate or what would be well-received by the students. How I wished for a book that would orient me in the subject and provide me with enough material for my classroom work! Since none was available, I struggled to develop my own resources. *Physics Is Fun!* was a result of my labors.

In the two decades since I first wrote this book, the new teacher's dilemma has changed: Now there are many books and manuals for Waldorf science teachers. In addition, the internet has become a pre-eminent resource. Instead of having too few resources, teachers now have too many, and it is difficult to choose among them.

I therefore decided to create a new edition of *Physics Is Fun!* to help teachers and home-school parents navigate the overabundance of choices for demonstrations and activities that could be presented to their students. I also wanted to share my decades of experience with a new generation of teachers.

In the first edition I strove to avoid being too directive, hoping that teachers would benefit from choosing among many options. I have heard from many teachers that they felt at a loss because there were too many choices. As a result, they ended up with a patchwork of experiences that didn't always result in clear concepts.

This edition is therefore much more carefully structured and directive than the previous one. I make specific suggestions for how to develop clear concepts by presenting a limited number of demonstrations and activities in a specific order. Teachers are, of course, free to find other resources and to make their own choices, but I can vouch for the effectiveness of my approach.

It's been more than thirty years since I was a new, young teacher facing the unknown. I have now taught physics to several hundred students and aspiring teachers and learned a great deal. I offer this new edition in the hope that all that I have learned will be useful to you.

Physics the Waldorf Way

1. The Purpose of Teaching Physics 1
2. Methods in Teaching Physics 5
3. The Structure of a Physics Main Lesson 9
4. Demonstrations, Recapitulations, and Reviews in Practice 15
5. Activities, Investigations, and Projects in Practice 21
6. Preparation and Materials for Teaching Physics 25
7. The Waldorf Physics Curriculum in Grades 6, 7, and 8 29

The Study of Physics in the Sixth Grade 35

Introduction to the Study of Sound 37

1. Listening 40
2. Producing Various Sounds 41
3. Mystery Sounds 42
4. Sounds Can Be Created by Blowing 43
5. Sounds Can Be Created by Plucking, Bowing, or Striking a String 43
6. Sounds Can Be Created by Striking 44
7. The Singing Glass 44
8. How Pitch Is Created in Blown Sounds, Part 1 45
9. How Pitch Is Created in Blown Sounds, Part 2 46
10. How Pitch Is Created in Plucked Sounds 47
11. How Pitch Is Created in Percussive Sounds 48
12. How Pitch Is Created in Sounds Created by Friction 49
13. How Sound Travels Through Different Media 50
14. Sounds at a Distance 51
15. Bath Time 51
16. How Sound Travels Through Solids 52
17. Cutlery String Chimes 53
18. Making a String Telephone 54
19. The Chladni Plate 55
20. Making a Bull Roarer 57
21. The Slide Whistle 58
22. A Jug Band 58
23. A Glass Harmonica Ensemble 59
24. Visible Sound 59
25. Recognizing Tone Quality, Part 1 60
26. Recognizing Tone Quality, Part 2 61

The Study of Light in the Sixth Grade 63
1. Total Darkness 66
2. The Unrevealed Revealer 68
2A. The Black Box. 69
3. The Straight and Narrow Path 70
4. Transparent, Translucent, Opaque 71
5. The Generation of Color 72
6. Creating Colors with a Water Prism .73
7. Working with Prisms. 74
8. Mixing Liquid Pigments 75
9. Mixing Light 76
10. Colored Afterimages 77
11. Afterimages from Afterimages 78
12. Colored Shadows 79
13. Orientation in the Darkness 81
14. Light and Distance. 82
15. The Intensity of Light 83
16. The Laws of the Shadow 83
17. Observing Shadows Cast by Sunlight. 84
18. Shadow Play 85
19. How Colors Look in Colored Light .85
20. The Bentham Disk 86
21. Making Simple Chromotropes 87

The Study of Heat in the Sixth Grade 89
1. The Sources of Heat: Combustion. 92
2. The Sources of Heat: Electricity. 93
3. The Sources of Heat: Chemical Reactions 93
4. The Sources of Heat: Friction 94
5. The Sources of Heat: Stress 94
6. The Sources of Heat: Light 95
7. Warming and Cooling. 95
8. The Ball and Ring 96
9. The Stretching Wire 97
10. The Expansion of Liquids. 98
11. The Expansion of Gases 98
12. Conduction in a Metal Rod 99
13. A Poor Conductor. 100
14. A Good Conductor 101
15. A Candle's Heat 101

16. Through a Glass Brightly 102
17. Visible Air Currents 103
18. Convection Currents in Water 103
19. Hooray for the Red, White, and Blue! 104
20. Warm to the Touch 105
21. Cold Hands 106
22. Cracking Marbles 107
23. Boiling Water in a Paper Cup 107
24. Hot Air Rises 108
25. Convection Tube 109

The Study of Magnetism in the Sixth Grade 111
1. Magnetic Magnetite 114
2. What Types of Metals Do Magnets Attract? 114
3. The Powerful Poles 115
4. Finding the North 116
5. Making Compasses 117
6. Like and Unlike 118
7. Push and Pull 119
8. A Magnetic Spring 119
9. Magnetism Travels Through Solids 120
10. Without Wetting Your Hands 121
11. No Visible Means of Support 121
12. Lines of Force 123
13. Revealing Attraction and Repulsion 124
14. Polar Power 125
15. Can Magnetism Be Screened? 126
16. The Three-Dimensional Field of Force 126

The Study of Electricity in the Sixth Grade 127
1. Amber's Properties of Attraction 130
2. Generating Static Electricity 130
3. Generating Static Electricity, Continued 131
4. The Attractive Power of Static Electricity #1 132
5. The Attractive Power of Static Electricity #2 132
6. The Attractive Power of Static Electricity #3 133
7. The Attractive Power of Static Electricity #4 133
8. The Attractive Power of Static Electricity #5 134
9. What Kinds of Materials Will Be Attracted to a Charged Object? 134
10. Attraction and Repulsion 135

11. Repulsion Between Like Charges 136
12. Reverse the Charges 137
13. Making an Electroscope 138
14. What Kinds of Objects Can Develop an Electrostatic Investigation? 138
15. Using an Electrophorus 139
16. Making a Leyden Jar 141

Teacher Resources 143

1. The Purpose of Teaching Physics

Why teach physics? Like every subject taught in Waldorf schools, the study of physics is a means to an end. What is that end? To nourish our students in soul and spirit and to help them develop into world citizens—people who know themselves, who understand and appreciate others, and who have a sense of responsibility towards the earth and all that live upon it.

To achieve these goals, teachers must strive to develop and strengthen the students' capacities and abilities to apply their knowledge and understanding in meaningful and responsible ways.

Developing the Student's Life of Feeling through Physics

In every subject we teach, we appeal to the students' feelings, for through the heart, the whole human being is engaged. A wise man once said, "All understanding begins with wonder." The least, as well as the most impressive, phenomena can evoke wonder if presented and received in the right way.

Although students may act as if they have seen, heard, and done it all, they can still be moved and awed by simple phenomena such as a candle shining in a dark room or the whirling pattern of convection currents rising and falling in a fish tank.

Our physics lessons should stimulate curiosity by confronting students with mysteries that are not always easily unraveled. The nature studies of the earlier grades were more directly comprehensible for students. They observed animals, found plants, or held minerals in their hands.

In physics in the upper grades, however, students will learn about forces that can be experienced only through their manifestations. As students learn more about physics, they develop concepts that are keys to understanding the natural world, the world of technology, and the nature of the human being.

As Rudolf Steiner said, we are increasingly surrounded by and dependent on the tools and products of modern technology, yet most of us do not understand the principles governing many machines, devices, and technological processes. By studying physics, students are helped to understand, appreciate, and accept the world that we have created. They learn to feel at home in the world and to approach their life tasks with confidence.

Developing the Student's Thinking through Physics

The upper elementary grades are a time of great change in the way students think. In the first three grades, students think primarily in imaginative pictures. As children journey through the fourth and fifth grades, their observations and experiences begin to transform imagination into thinking.

By the time they enter the sixth grade, students are beginning to be capable of developing other capacities of thought: causal, conceptual, analytical, and deductive thinking. The study of physics addresses and develops these new capacities.

The phenomena of physics can only be understood in causal terms because we cannot see the forces at work. Pluck the string of a violin and a sound arises, loud at first, then becoming softer. The string vibrates strongly at first, then its vibration decreases. What is the relationship between the vibrations of the string and the sound? The vibrations can be seen, the sound heard. Is sound vibration? Has it been caused by the vibration? Such questions are central to the study of physics, and their answers demand that the students apply new capacities of thinking.

In our teaching of physics, experiences must be raised to the conceptual realm, for without concepts, physics remains a potpourri of experiences, interesting and even memorable, but not necessarily meaningful. Students in the upper grades of the elementary school begin to become more objective. The capacity for objectivity develops slowly, and only after the age of twelve are students able to divorce themselves sufficiently from their own experiences to reflect upon them, to use their senses more consciously, and to analyze their perceptions.

Forming concepts is a mere intellectual exercise if it does not lead to a deeper understanding of nature and a fuller appreciation of technology. Throughout every stage of their physics studies, students make connections between the subjects presented in the classroom and the rest of life. Science lives not in the laboratory but in the world. While the scientist may discover some of nature's secrets, any informed person should be able to read, understand, and appreciate the book of nature. At the end of their physics studies, students should feel more comfortable, competent, and confident—better able to meet the world's challenges and to solve problems creatively and effectively.

Developing the Student's Will through Physics

As we have seen, the study of physics develops and engages the student's life of feeling and develops the new powers of thinking of the young adolescent. It also promotes the healthy development of the will, of purposeful intention and action.

The way we teach physics develops the students' wills. By observing the teacher's orderly, methodical demonstration technique, the students are helped to form clear and orderly thoughts regarding the phenomena they are experiencing. When the demonstration is reviewed the next day, the students recall what occurred, they account for the materials and procedures, and they formulate the concept. Through such daily work the students' memories are strengthened, and their thinking is infused with will.

When the students engage in physics activities or investigations, their will is further developed. The work of the scientist is logical and orderly. Although some great discoveries have occurred by accident, most scientific exploration requires scientists to follow specific procedures and to account precisely for what has been done. As the students learn to plan their activities, formulate their ideas, and use laboratory equipment, they are strengthened in their capacity to work with purpose.

A Waldorf lesson is not complete unless there is plenty of time for the students to do their own academic, artistic, and practical work. This is especially true in the physics main lesson blocks, for the bulk of the written work should come from the students themselves. The teacher may help the students formulate a clear statement of the concept, but that concept is more meaningful when students write about or illustrate their experiences and ideas in their own words and images.

Finally, physics awakens the students' moral forces. Like all subjects in the Waldorf school, physics places the human being at its center. As students begin to perceive the physical world in a new way, they also learn how the structure of their bodies expresses physical laws—their eyes function like cameras, their legs and arms like levers.

As they begin to understand how the laws of nature manifest through physical forces, they also learn how humans have harnessed and channeled these forces through technology. Technology may be used for constructive and destructive ends. In their physics lessons, students learn to work with materials and equipment in a purposeful and constructive manner. This helps them to recognize that the earth's problems can only be addressed by developing a sense of moral responsibility towards the material world.

The teaching of physics strives for the same goal as all other subjects: the development and cultivation of the students' thinking, feeling, and willing. We must keep this goal firmly in mind and not let the subject matter assume an overwhelming importance.

The content of our lessons is a means to an end—not the end in itself. Let us not forget that the essence of our work is what we do with the students. What occurs in them, what they are able to develop through our lessons, is much more important than the topics they study.

2. Methods in Teaching Physics

Waldorf education does not provide formulas for teachers to follow; rather, it demands that teachers develop pedagogical practices out of an anthroposophical understanding of the human being. The foundations for this understanding were laid by Rudolf Steiner, but teachers must work to make it their own. Through study, practice, and reflection upon our experiences, we can derive a method that truly addresses the whole human being.

The Foundations of the Waldorf Method

In Lecture 9 of *The Foundations of Human Experience*, Rudolf Steiner analyzed and described the process of human thinking. He wanted teachers to develop a clear understanding of this process since they were going to be particularly responsible for developing their students' intellects. In Lectures 2 and 3 of *Waldorf Education for Adolescence*, he reviewed some of these principles and showed how they applied to the organization of the lesson.

According to ·Rudolf Steiner, thinking occurs in three stages, which are interwoven and often occur simultaneously. He identified these stages as drawing conclusions, forming judgments, and arriving at concepts.

Rudolf Steiner's View of the Human Being

Rudolf Steiner articulated a view of the human being that has both sensible and supersensible "members." In this section, reference will be made to the *ego*—that imperishable part of our true self; the *astral body*—our "feeling body," which allows us to have sensations, perceptions, and emotions; the *etheric body*—our "life body," which is responsible for all of our life functions and also for memory; and our *physical body*—the mineral basis for our physical existence. For those who are interested in learning more, see Rudolf Steiner's extensive descriptions in *Theosophy* or in *An Outline of Esoteric Science*.

Drawing Conclusions

In Rudolf Steiner's view, conclusions are perceptions brought to consciousness. Any sensory impression can become a conclusion, but only if we first become aware of it.

We often have impressions of which we are not aware. For instance, we might hear—but not listen to—a sound. Hearing consists of perceiving sounds, while listening requires our active involvement. We all have experienced being in a crowded room and hearing many conversations, but if someone mentions our name, we hear that word emerge from the cacophony of sound. In order to draw conclusions we must, so to speak, listen to what we hear. We therefore draw conclusions about only some of our perceptions.

We continually live in the realm of conclusions. When we speak we express conclusions; when we listen to the speech of others, we draw conclusions; much of what we read remains in the form of conclusions. According to Rudolf Steiner, conclusions can only live in the soul when it is awake. They are a result of the working of the *ego*, which helps us make sense of the world. While conclusions affect the whole human being, they particularly affect the life of thought, for they serve as foundation for our judgments and concepts.

One of the main tasks of the teacher is to present the students with conclusions that can become the foundation for concepts. If, instead, we present the concepts directly, they will live in students' souls as conclusions that will not be capable of growing and developing further.

Forming Judgments

When we become conscious of our perceptions, we begin to form judgments. According to Rudolf Steiner, this process arises out of the human being's desire to transform the process of perception into knowledge. A conclusion consists of the awareness of our experience, but a judgment allows us to know that experience more fully and to relate to it. Our conclusions live in our sensory impressions; our judgments allow us to "make sense of" those impressions. We do this by weighing our perception against our prior experiences and our memories of similar percepts from the past.

Judgments live in the dreaming soul, so we are not fully aware of them. They are a result of the work of what Rudolf Steiner called the *astral body*, that part of ourselves that allows us to have perceptions, feelings, and emotions. While the drawing of conclusions only affects the way a person thinks, the manner in which a person learns to form judgments affects her life of habits, her very character. Thus we teachers have a special responsibility to help students form judgments in a sound and healthy way.

Arriving at Concepts

Conclusions and judgments are inextricably linked. There is a constant interplay between our awareness of a percept and the process of linking it with our prior experience. A conclusion is complete as soon as we become aware of it. The finality of a conclusion is well expressed in its German form, *Schluß*, which means "finished" or "done." The process of forming judgments is not over so quickly. We can continue to ponder and weigh our impressions for a while, but at some point, this process also must come to an end.

Concepts, on the other hand, grow gradually in the human being and must never be finished or fixed. It is the teacher's role to help the students form concepts in such a way that the concepts can grow and become increasingly well developed and refined. Conclusions develop awareness, judgments result in knowledge, but concepts form the basis for wisdom, allowing us to understand ourselves and the world around us.

Concepts live in the sleeping soul, and unless we school our thinking, we will not become aware of them at all. They are a result of the working of what Rudolf Steiner called the *etheric body*, that part of ourselves responsible for life functions and also for memory. The concepts that we form have an effect on the physical body. According to Rudolf Steiner, a mature person's features and physiognomy reveal the types of concepts that were formed early in life. This places a very special responsibility on the teacher to strive always to engender living concepts in the students.

While fixed or dead concepts result in the hardening of the soul, living concepts allow a person to remain mobile and flexible in her thinking, and they provide the basis for inner freedom. What does it mean for a concept to grow? Let us take as an example a concept that is developed throughout the grades—the concept of light. Here are some of the many ways that children develop a relationship with this concept: In the earliest school years, children experience light and they experience darkness. Their experiences are not brought to consciousness, but we know that these experiences make a deep impression on their souls. In the early grades, children recite poems and sing songs that describe or express the light. They say the morning verse that begins, "The sun with loving light..." They hear stories and legends involving light and the darkness. Through these experiences they form moral associations that will remain rooted deep in the soul.

In the third, fourth, and fifth grades, students are told about the creation of light in the stories from the Hebrew Bible and various mythologies. They also learn about light in their studies of the animals and plants. In their farming and in geography classes, they come to understand the effect of light on crops, animals, and people. In the sixth-, seventh-, and eighth-grade study of physics, students begin to study light in all of its manifestations. They learn the characteristics of light, its behavior, its laws. Now the concepts they form can be expressed in concise and concrete terms.

Throughout the high school years, students return again and again to the study of light. In astronomy and meteorology, in physics and chemistry, in economics and philosophy, they come to understand and express the many aspects of light. Through these years the concept of light grows ever more complex and also ever more refined. By developing concepts in this way, we help students to connect to the world and to trust that they can understand it ever more fully.

3. The Structure of a Physics Main Lesson

In Lecture 3 of *Waldorf Education for Adolescents*, Rudolf Steiner indicated how physics lessons can be presented in a two-day rhythm to engage the whole human being. Such a schedule consists of a demonstration and a recapitulation on the first day and a review on the second day. This allows students to draw conclusions, form judgments, and arrive at concepts in a living way.

The Demonstration

When the demonstration is performed, students draw a series of conclusions from their observations and impressions. The demonstration acts powerfully, affecting the whole being of the student and requiring the strength of the *ego* to bring perceptions to consciousness.

The Recapitulation

Immediately after the demonstration, the teacher verbally recapitulates the demonstration for the students, briefly describing what was done and what occurred. This recapitulation allows the students to re-experience through their memory and imagination the events they observed with their senses. The experiences that took hold of the whole body are now lifted into the rhythmic system and involve the *astral body*. The recapitulation promotes the forming of judgments and will provide the basis for living concepts. After the recapitulation, no further work is done that day with these phenomena.

During sleep, the *ego* and *astral body* leave the physical and *etheric body* and enter the spirit realm. The impressions of the demonstration and the pictures that the students formed during the recapitulation are then lifted into higher realms. When the students awaken the next day, these impressions and pictures are present again, but they have been transformed. They now live in the *etheric body* as memory pictures and as the germs of living concepts.

The Review

The next day, when the students return to school, the first part of the lesson includes a review and discussion of the previous day's work. Now the students summarize their observations and articulate their judgments. New judgments may be formed as students listen to each others' experiences and hear about aspects of the demonstration that they may not have observed or remembered.

As students refine their judgments and come to a deeper understanding of what they observed, concepts begin to emerge. Because the concepts are still in a germinal form, it is preferable to frame or phrase them in general terms. Articulating a concept too narrowly causes it to become fixed in the students' minds and deprives it of some of its power to grow and develop.

Once the concept has been stated—perhaps differently by different students—and then summarized by the teacher, the students express the concept in their written and artistic work. In the further demonstrations, they will be able to build upon or develop the concept further and to recognize its manifestations and applications in their lives and in the world.

The three parts of the main lesson correspond to the three processes of thought and address the threefold human being. During the demonstration, students primarily draw conclusions about the phenomena. The process of drawing conclusions takes hold of the whole human being, calls strongly upon the students' *egos*, and strengthens the will.

The recapitulation directly after the demonstration calls on the students to form judgments and helps them take the content of the demonstration into their sleep life. The recapitulation lifts the experience of the demonstration into the rhythmic system and engages the *astral body*, nourishing the students' feelings.

The review on the next day helps the students begin to arrive at the concept. Calling on the forces of memory, the review engages the *etheric body*, developing the students' thinking.

PART OF LESSON	THOUGHT PROCESS	SPIRITUAL MEMBER	SOUL FACULTY	BODILY FOUNDATION
demonstration	conclusion	ego	willing	legs/feet
recapitulation	judgment	astral body	feeling	chest/arms
review	concept	etheric body	thinking	head

If we extrapolate this two-day rhythm into the course of a week, each day contains all three parts of the lesson: review, demonstration, and recapitulation. Each day also includes a significant amount of time for the students to do their academic and artistic work. While the latter parts of main lesson may be organized in several ways, it is essential that the review of the previous day's work come first, right after the warm-up exercises.

In my lessons, the review is typically followed by the written and artistic work arising out of the previous day's topics, and the new demonstrations or activities come last. This allows the new phenomena to be the last thing students experience in the main lesson.

Some teachers prefer to present the new material after the review and finish the lesson with a work time. This works best if the new material follows directly on what was reviewed.

I recommend the following schedule:

1. Warm-up exercises: 15–20 minutes (in seventh and eighth grade, this can be shorter).
2. Review of previous day's work: 15–20 minutes (in seventh and eighth grade, this can be longer).
3. Written and artistic work: 50–60 minutes.
4. Demonstration and recapitulation: 20 minutes.

(If students are performing an activity or investigation, this last part of the lesson will need to be extended and the work time reduced.)

It is tempting to let the warm-up exercises and the review go on too long, thereby shortening the time for working in the main lesson book, which will contain all of the the student's compositions, lab reports, creative writing, illustrations and diagrams relating to the course. We need to commit to giving students sufficient time to do their individual work every day. Too often this part of the lesson is curtailed, and the students' wills are not sufficiently engaged.

Experiencing Phenomena through Demonstrations, Activities, Investigations, and Projects

In Waldorf schools, students experience and learn about physical phenomena in four ways: demonstrations, activities, investigations, and projects. The Waldorf approach to teaching physics allows the phenomena revealed by the demonstrations, activities, investigations, and projects to speak directly to students without intermediary concepts or explanations.

Demonstrations

Demonstrations should be used to introduce phenomena; their goal is to intrigue and reveal, to tantalize and inspire. By engendering wonder and stimulating curiosity, demonstrations work directly on the students' feelings, priming them for further learning.

A demonstration should need no explanation; it should be sufficient unto itself. It should allow a phenomenon to speak without words directly to the students. It should be a memorable experience, imprinting itself on the students' minds and hearts as a manifestation of the topic they are studying.

Note: Teachers who succumb to the temptation to demonstrate phenomena as if they were magic tricks generally find that they have stimulated their students but have not established an adequate foundation for conceptual understanding. For that reason, I encourage teachers to keep demonstrations simple.

For demonstrations to work properly, both teachers and students need to understand and accept their roles. The teacher's role is to perform the demonstration, the student's is to observe. This is not the time for explaining or for asking and answering questions. Anything that distracts the students from perceiving the phenomenon compromises the demonstration and interferes with the student's direct perception of natural laws from which true understanding will grow.

The demonstration approach was considered the most effective method of science education for much of the nineteenth and twentieth centuries. This approach fell out of vogue during the second half of the twentieth century because it was thought that demonstrations do not involve students actively enough. I have found, however, that demonstrations *do* involve students actively, but on an inner, rather than an outer, level. This inner experience provides a sound foundation for students' artistic and academic work with the subject

Teachers who perform demonstrations will recognize that they deepen their students' level of appreciation and understanding. While it may appear that demonstrations engender passivity and a blind acceptance of the teacher's scientific authority, the opposite actually occurs. Students who have experienced good demonstrations typically have more questions and are more eager to understand the phenomena than those who learn through other types of activities.

Activities

When students engage in activities, they study phenomena in a direct and vivid way. They develop a practical knowledge about how the phenomena work. This grounding in physical reality serves as a firm foundation for conceptual understanding.

The teacher's demonstrations present the general principles of physics, while activities draw the students into the realm of the particulars, the details. Details will only be meaningful, however, if they relate to a larger whole. If we wish our students to gain a conceptual understanding of the subjects, we must plan activities that are purposeful, so that they build on each other to reveal greater truths.

Activities should address the question *how* on the most practical and empirical level. In order for this question to be answered thoroughly, students should explore a subject in various ways. Activities should be organized so that students will gain a personal experience and understanding of the principles that govern the phenomena.

By working through a series of activities, students come to the point where they really know the topic. They can look beyond the particulars to the general principle; they can express the concept.

At this point, they may be ready to research a particular aspect of the topic and write reports; now they can formulate meaningful questions that can be answered by further investigation.

Investigations

Demonstrations and activities engender wonder, stimulate curiosity, and spark interest. They reveal the most obvious manifestations of a phenomenon, but the underlying laws and practical applications are waiting to be discovered. If a demonstration basically addresses the *what* of a phenomenon, and an activity the *how*, the investigation is designed to answer the question *why*.

Topics for investigation will often arise out of the students' questions. These topics need to be carefully developed, however, to be on a level that will give meaningful results. An ambitious or sophisticated program of scientific investigation may intimidate the students who are not able to analyze the data sufficiently, but an investigation that is too simple may not sustain the students' interest.

Investigating a phenomenon should be a serious endeavor and should not degenerate into playing around with materials and techniques. An investigation should be clearly designed so that its results answer a question that has previously been formulated. Just trying things out and seeing what happens may be interesting and fun, but such an approach does not give students the sense of scientific experimentation as a purposeful and orderly process.

Investigations that are well designed will be satisfying and will stimulate the student to develop new levels of understanding. Investigating a phenomenon gives students the sense that their work is important; it engenders self-respect as well as a heightened respect for those who devote their lives to the exacting and challenging work of scientific discovery.

If teachers develop a sense of the deeper purpose and the inner qualities of the different parts of the physics lesson and its activities, they will teach in a way that engages the whole human being. Through such physics lessons, the students' feelings will be awakened, their thinking enriched, and their wills developed in a balanced way.

4. Demonstrations, Recapitulations, and Reviews in Practice

Demonstrations

For demonstrations to be well received and well understood, they need to be well planned and well performed. Here are the practical steps we need to take to accomplishing our goals:

Preparing Demonstrations

1. Keep the characteristic you are planning to demonstrate clearly in mind.
2. Gather all the materials for the demonstration. Materials should be familiar objects that are large enough to be seen clearly.
3. Arrange all the materials in the order they will be used so that you do not have to search for an object.
4. Rehearse the demonstration several times so that it can be done fluently with no unnecessary movements.
5. When rehearsing, observe the demonstration carefully.

Performing Demonstrations

1. Organize all the materials in the order in which they will be used.
2. Arrange the students in the classroom so that they can observe the demonstration without distraction.
3. Remind the students that they should be quiet and attentive; there should be no questions or comments during the demonstration.
4. Introduce the demonstration by identifying the subject or topic to be demonstrated, e.g., "Today we will observe how light travels." Do not state the characteristic that the demonstration will reveal.
5. Perform the demonstration slowly enough so that all students can observe the phenomenon. When necessary, direct their observation to an aspect of the phenomenon they might not notice.
6. Do not explain what you are doing or why the results occur.

Recapitulating

A demonstration will sink in most deeply if it is recapitulated and reviewed. The recapitulation is done by the teacher directly after the demonstration. During the recapitulation, students listen to the teacher; they don't share their own observations or perceptions. Some teachers may initially feel awkward telling the students about what they have just experienced. One way to overcome this is to introduce the recapitulation by saying, "Now try to recall in your mind's eye what you have just experienced."

1. When the demonstration is over, students should return to their seats. Briefly recapitulate the demonstration in words. Describe what was done and what occurred so that the students can re-experience it in their memory and imagination.
2. An example of a recapitulation might be, "Let's think back on what we've done today: A copper rod that had small balls of wax attached at one-inch intervals on its underside was supported on a stand. One end of the rod was heated with a Bunsen burner. After a few moments, the wax ball nearest the flame fell off; then other balls began to fall off in succession. The last two balls did not fall off even after the rod had been heated for several minutes. Tomorrow we will go on to observe how heat travels in liquids and gases."
3. Put the materials away and go on to the next part of the class.

Reviewing

1. On the next day, review the demonstration thoroughly through a series of questions that can be answered orally and/or in writing.
2. Reviews should account for the topic, the materials, the procedures, the observations, and the results. They should lead to a clear formulation of the characteristic shown by the demonstration and, where possible, the concept.
3. While reviews should be methodical and logical, they need not cover everything. If the reviews are too similar day after day, they will become boring, and students will lose interest. Reviews will be different at the different age levels. Seventh and eighth graders will be much more eager to interpret the results, while sixth graders will be challenged to recall what occurred and will want to share what made the deepest impression on them.

SUGGESTIONS FOR REVIEWING

1. *Synthetic approach*: What was done first, second, third, etc.? What occurred first, second, third, etc.?
2. *Associative approach*: What happened? Where else have you experienced this in your life?
3. *Affective approach*: What did you like best? What impressed you the most?
4. *Predictive approach*: What happened? What did you think would happen?
5. *Analytical approach*: What happened? Why did it happen?
6. *Comparative approach*: What happened? How does this compare or contrast with what happened in other instances?
7. *Hypothetical approach*: What do you think would happen in the following hypothetical situation?

These approaches may be used singly or in combination, but it is best not to try to work in too many ways at any one time.

Note: A wise colleague once said to me that there are three kinds of questions that should arise in a review: questions to which you know the answers; questions to which you don't know the answers; and questions which you never thought to ask. We should try to resist the temptation to over-structure the review and ask too many factual, closed-ended questions, as if information were the key to meaning. If we trust that what is important will emerge, our trust will be rewarded.

Written Work

Physics is an excellent subject for the practice of clear expository writing. In sixth grade you can still sometimes have the students copy summaries from the board or dictate them, but in seventh and eighth grades the bulk of the writing should be composed by the students.

If the students have to summarize every demonstration, activity, or investigation, they will begin to resist recording their experiences. It is therefore important that the written work be economical and varied, with plenty of opportunities for creativity. While it may be tempting to use the standard lab report format exclusively, I think that this format is limited and should be used sparingly.

SUGGESTIONS FOR WRITTEN WORK

1. A clear *expository account* of the demonstration
2. A step-by-step *recipe* of how to achieve the demonstration's results
3. The standard *lab report*, which lists materials, describes procedures, records observations, and states conclusions
4. A *list or chart* of aspects of physical forces, e.g., *Sources of Heat* or *Uses of Electricity*
5. A *poem* about the phenomenon
6. A *story or play* whose theme is derived from the phenomenon
7. An *opinion essay* regarding the uses of a physical force, e.g., "Electricity—Servant or Master of Humanity?"

Artistic Work

The artistic work that accompanies, illustrates, or illuminates the written work is just as important in the sciences as in the humanities, and students should do some sort of artistic work every day in the main lesson. Different kinds of artistic work have different effects on the students, and we will need to determine when paintings, drawings, technical illustrations, diagrams, or charts are most appropriate. Paintings express the mood of a subject, while drawings convey the setting or context. A technical illustration focuses on the materials, while a diagram shows the inner structure of a particular object or process. A progression that leads from the imaginative to the structural should be utilized consciously, depending on the subject matter.

SUGGESTIONS FOR ARTISTIC WORK

1. Decorated or illustrative titles
2. Illustrative or practical page borders
3. Diagrams of the equipment
4. Illustrations of the phenomenon in the classroom
5. Graphic or cartoon renditions of the phenomenon
6. Drawings of the phenomenon in nature
7. Paintings

A Postscript on Demonstrations that 'Don't Work'

Sometimes a demonstration does not yield the results we expected. Does that mean that it didn't work? I do not think so. Natural forces manifest in a consistent manner, though we may not always perceive or understand them clearly. If a demonstration does not work the way we thought it would, we need to observe it more carefully and perceive more clearly what is happening so that we might begin to understand the causes of the phenomena we are observing.

What should we do if things don't work according to our expectations? We might a) proceed; b) try again; or c) transform the demonstration into an investigation.

Proceeding with the demonstration as if what is happening is exactly what should be happening is true to our process. We are seeking to know how the forces of nature behave, not to control them. If these forces behave in unexpected ways, we have a special opportunity to learn from them. If we become too fixed in our notions of "what should happen," we may actually be limiting the way that we view the phenomena of nature.

If the results of our demonstration are unexpected, it is appropriate to say to the students: "This demonstration did not proceed as I expected. I am going to repeat it so that we can all observe it more carefully and try to understand what is happening." This will alert the students to the fact that something mysterious has happened and that they may help solve the mystery. During the review on the following day, the teacher might describe what she expected would happen and see if the students can figure out the causes for what did happen.

If a demonstration "doesn't work" because we have not prepared sufficiently or have performed it incorrectly, we have an opportunity to show students how we deal with our shortcomings. As teachers, we are primarily learners, and mistakes are some of our best learning and teaching opportunities. How we react and how we work with such situations affect the students much more deeply than the details of a demonstration that "didn't work."

5. Activities, Investigations, and Projects in Practice

Activities

Activities differ from demonstrations in that the students are more directly involved. In demonstrations, students observe; in activities, they do.

Preparation

1. Identify the subject for the activity. The characteristic or underlying concept should stem either from a previous demonstration or from the students' experience in a subject. Students may be asked to offer suggestions for various ways to work with a subject.
2. Choose the activities that will best illustrate the phenomenon. An activity should be fairly simple to perform, be interesting to observe, and give clear results.
3. Gather all the materials for the activity. Students can help determine what will be needed and can gather some of the materials.
4. Arrange all the materials and tools conveniently so that the students have what they will need and don't have to search for anything.

The Activity

1. Arrange the students individually or in groups of no more than four, and set up the room so that everyone can work with concentration on the activity and observe without distraction.
2. Remind the students about the purpose of the activity. If appropriate, appoint—or let each group choose—someone to record the results.
3. As the students perform the activity, keep talking to a minimum so that everyone can focus on the phenomenon.
4. When necessary, direct the students' observation to an aspect of the phenomenon they might not notice. Do not explain or interpret what is being done or why the results occur.

5. When the activity is over, have students put the materials away, or wait to clean up until after the recapitulation.
6. Like demonstrations, activities should be recapitulated, reviewed, written about, and illustrated. Students might also research the topic further and write reports.

Investigations

Investigations involve students on an individual level, and they need to be planned carefully so that students can perform them successfully. Investigations will be most meaningful if they arise out of the questions that came up in a demonstration or activity. Investigations should be logically organized, be reasonably simple to perform, and yield results that are clear and easily interpreted.

Preparation

While students will conduct the investigation themselves, they still need guidance from the teacher.

1. Narrow the topic for investigation and have the students identify and formulate what aspects of a topic they plan to investigate. The investigation should be stated in terms of a question, e.g., "Do all metals conduct heat equally well?"
2. Have students make a list of materials that will be needed for the investigation.
3. Students should state the procedures they will follow to perform and record the investigation.

The Investigation

1. If the investigation is being done in class, students should work where they can focus and not distract each other.
2. Students should observe the investigation carefully and record the results accurately.
3. Students should repeat the investigation at least once to verify the results. During the review on the following day, students should describe the investigation and its results.
4. Specific answers to the original questions should be given and, if possible, explained.
5. Students should write about their investigations, clearly stating the topic, procedures, observations, results, and conclusions.

6. Related investigations of various aspects of a topic should be discussed and compared.
7. Further topics for investigation may be suggested and should be formulated as questions.

Projects

The work that we do in the classroom with the entire class can be expanded and deepened by individual or group projects. Many demonstrations and activities can be prepared by students and then presented to the class. In addition, students can render their work in many forms besides the types of assignments that have been suggested.

In sixth and seventh grades, projects might be small in scope, but in the eighth grade, I often assign a major project that students work on throughout the physics block. During the last week of the block they present their projects to the class.

Ideas for Projects

1. Research and write a report about a topic.
2. Interview someone about the topic and present the interview using an appropriate medium.
4. Create a set of illustrations, a poster, photomontage, or collage.
5. Produce a video about the topic.
6. Create a slideshow or multi-media presentation.
7. Produce a demonstration or activity manual.
8. Write a children's story about the topic.
9. Create an illustrated, graphic, or cartoon account.
10. Compose a humorous skit or sketch illustrating the topic.

6. Preparation and Materials for Teaching Physics

Preparation for Teaching Physics

Class teachers do not need to become scientists or specialists; rather, they need to become sufficiently well-grounded in the subject to introduce and develop it in a way that gives the students a sense of its breadth and depth.

As we have seen, in their studies the students draw conclusions, form judgments, and arrive at concepts. Teachers go through the same process in their preparation. This reinforces the importance of the preparation. By experiencing the phenomenon before introducing it to students, a teacher draws conclusions. By connecting the new experience with prior ones, a teacher forms judgments. By contemplating the physical laws implicit in the phenomenon, the teacher develops new concepts or enriches existing ones. This process is especially meaningful when teachers prepare to teach physics for the first time.

Preparation Schedule

Teachers who do not prepare enough in advance find themselves desperately trying to "bone up" on the subject on the weekend or evening prior to the main lesson. In such cases, the material tends to remain undigested and will probably not be presented to the students in the clearest and most comprehensible way.

The first stage of preparation should take place during the year prior to teaching physics for the first time. Such long-range preparation allows our ideas and concepts about the subject to grow and mature in a healthy and organic way. It allows us to awaken to the subject, to develop a relationship to the subject, and to formulate questions.

The second stage of preparation should be done the summer prior to teaching. This preparation consists of gaining a basic grounding in the subject by reading about it in encyclopedias, reference books, websites, and trade books. Unless one has a special interest, the class teacher's basic grounding for teaching physics should remain on a layperson's level; one does not have to become a physicist. The summaries contained in children's encyclopedias are especially useful, for they are succinct, clear, and easy to understand. Using these as a foundation, we can read further, knowing what to look for.

This kind of background reading is best done the summer before teaching physics. Strange as it may sound, it is important for most of the facts to recede into the background, and if we allow enough time, this happens naturally. What will be important in the actual teaching of the main lesson block is not the teacher's intellectual grasp of the subject, but her relationship to the subject. This relationship needs time to grow.

The summer before the next year of teaching is also the best time to use online resources. The internet is a treasure trove of information, but the multitude of demonstrations and activities makes it hard to find what is truly useful. I have discovered some compelling ideas for demonstrations online, but often I stumbled upon them. That kind of discovery process is best done in the summer, when one is still in a browsing mode. I caution against depending on the internet for help shortly before teaching a topic because it is likely to offer too many choices and will lead to confusion rather than clarity.

During the summer before teaching the main lesson block, we should try to gather the materials and try out the demonstrations, activities, and investigations. In the summer we have the time to try out several ways of demonstrating a phenomenon and to determine which way is best. We have the opportunity truly to ponder the phenomena that are being introduced, and, in the following months, to find other ways these phenomena reveal themselves in nature and in everyday life. Finally, if we begin working with the phenomena during the summer, we will not look forward to the physics block with dread and fear, but with eager anticipation.

On the weekend prior to each week of physics teaching, we should go through all the demonstrations, activities, and investigations once more. Although this sounds like a lengthy process, it does not take more than about an hour if all the materials have been gathered before the block begins.

Doing all of the week's demonstrations, activities, and investigations at one time allows us to see whether all that will be presented forms an integral whole and whether the concepts we are trying to engender can be perceived from the sequence we have chosen. Doing all of this on the weekend before is a sure remedy for panic and dread; we can then approach the week confident that our classes will go smoothly.

Finally, it is worthwhile to go through each demonstration or activity one last time on the afternoon prior to the morning that it will be presented. This could also be done early that morning, but an element of tension might arise that is not welcome or conducive to good teaching. This will be the fourth or fifth time that we will have dealt with the phenomenon. By now the materials and the procedures will be well known; we can therefore perform a demonstration smoothly and effortlessly. There is little or no chance that something will go wrong, for every move will be well practiced.

SUMMARY OF PREPARATION SCHEDULE

Year before	Orient ourselves to the subjects.
	Observe phenomena; find connections to subjects.
Summer before	Research the background of next year's subjects.
	Gather equipment and materials.
	Try out demonstrations and activities.
Vacation before	Choose the specific subjects.
	Gather the materials for demos and activities.
Weekend before	Run through the next week's demonstrations.
Afternoon before	Do the next day's demonstrations.

The kind of preparation described above requires a commitment on the part of the teacher to form a relationship to the subject, to become proficient in the subject, and to be well-prepared on a daily basis. Although I have taught physics for more than 35 years and have performed most of the demonstrations many dozens of times, I still prepare for my teaching in the way I have described. Such preparation allows me to focus on the phenomena and not on what I am doing in my lessons and to be more aware of what my students are experiencing as they encounter the mysteries of the physical world.

Materials

Teaching physics is a wonderful opportunity to lead the students into a new relationship with the material world. Because we rely so much on physical materials, physics is a subject that allows students to form a new conception of man and matter.

The materials we use change as the curriculum changes. In sixth grade, we should try to use everyday objects whenever possible. The teaching of physics in the sixth grade helps students become aware of their sense perceptions, of the natural world, and of the physical objects in their lives. By using everyday objects we develop this awareness out of daily life itself. Were we to use sophisticated equipment at this time, students might begin to feel divorced from the phenomena and start to think that science happens in the laboratory rather than all around them.

By using everyday objects we also allow students to work further with the topics we have explored. If they can readily obtain these materials, students will try the same demonstrations and activities at home, thereby deepening and reinforcing the experience they gain in the classroom. Using expensive or complex equipment precludes the possibility of students integrating their school activities into their daily lives.

In the seventh and eighth grades, there are occasions where everyday objects are not sufficient. As students begin to measure and calibrate more carefully, they may need to use more sophisticated measuring tools. As they begin to explore phenomena more specifically, they may need to use some laboratory equipment. I have met some teachers who reject such equipment on philosophical grounds. This can frustrate the students and instill the feeling that they are not actually working scientifically. On the other hand, we can easily forget that what makes science "scientific" is our approach, not the tools we use. In a society that places such a value on apparatus and technology, we should remain clear that the process of discovery is not dependent on our equipment, but on ourselves.

I have found that the physics departments in local colleges and universities are often willing to lend equipment so that a school does not necessarily have to invest in expensive items such as an electrostatic generator or a vacuum pump.

7. The Waldorf Physics Curriculum in Grades 6, 7, and 8

The physics curriculum of the upper elementary grades is rich and varied. In the course of these three years, the students are introduced to all the major areas of physics. Rudolf Steiner gave the following indications regarding the teaching of physics in the *Curriculum Lectures* of September 6, 1919:

> *We start with the teaching of physics in the sixth class in such a way that it is linked with what the child has acquired through the teaching of music. We start our teaching of physics by letting acoustics grow out of the musical. You bring acoustics into relation with musical science when you describe the human larynx physically and physiologically. You cannot describe the human eye yet at this age, but you can the human larynx.*
>
> *Then you proceed to optics and the science of heat, taking the most important items. The principles of electricity should be taught in the sixth class as well.*
>
> *In the seventh class you extend the teaching of acoustics, heat, optics, electricity, and magnetism, and only from here do you proceed to the most important principles of mechanics, i.e. the lever, the wheel and axle, roller, pulley, inclined plane, cylinder, screw, etc.*
>
> *In the eighth class you enlarge again, revise what you have fostered in the sixth class, and then pass on to hydraulics, the theory of power working through water. You take everything belonging to the conception of lateral pressure of the water, buoyancy, the Archimedian principle, etc. ... You bring the physics lessons to some sort of a conclusion with aeromechanics whereby everything is discussed concerning climatology, the barometer, and meteorology.*

Such a curriculum sounds formidable. How is it possible to cover all of these subjects in one main lesson block of three or four weeks each year?

Some teachers have solved this problem by increasing the number of weeks devoted to physics during the year. Others have coped by reducing the number of subjects covered in a year. In sixth grade, for instance, only sound, light, and heat might be introduced; in seventh grade, only mechanics and electricity, and in eighth grade, only fluid mechanics. This approach seems to have the advantage that the topics can be covered more thoroughly and in greater depth. To do so is to vary from what Rudolf Steiner suggested, however, and I urge you to try to follow the sequence outlined in the quote above.

One of the reasons why we should include all the subjects each year is because students at the different grade levels relate to the topics differently, and each grade benefits from a different approach. Sixth graders who are studying physics for the first time come to the subject with tremendous enthusiasm. Every phenomenon makes an impression. Because they are so open to the world, everything is new, exciting, and interesting. Because their sensory impressions are so strong and compelling, sixth graders are fully engaged by phenomena. If we help them become more aware of their sense impressions, the conclusions they draw will be sound and true.

Seventh graders approach the study of physics in a different way. The subject is no longer exciting because of its newness, but it is more interesting and compelling because of the beauty and truth of its laws. Most seventh graders are no longer content simply to observe the phenomena; they want to be more involved and to see how topics relate to them. The approach to physics in the seventh grade must therefore change to meet this urge for involvement, and we must find ways in which students can become more active.

Seventh graders are also eager to quantify their experience, and the physics work should satisfy this desire by incorporating measurement wherever possible. The task of the seventh-grade teacher is to help the students reflect upon and objectify their experiences and to form sound judgments.

By eighth grade, students want to be treated like young adults, and they might balk at a demonstrational approach because they consider it babyish. Because eighth graders need to experience the relevance of what they are studying, we need to try to incorporate the larger spheres of life and industry into all that we study. Students still need to observe accurately, however, so that they can draw clear conclusions. They need to measure carefully, so that they can form sound judgments, but now they need to be able to take initiative and try things out for themselves. Out of this activity, concepts should grow—concepts that are rooted in observation and weighed against experience; concepts whose application may be seen working in the human being, in society, and in the natural world.

The process of thinking—drawing conclusions, forming judgments, and arriving at concepts—is developed in every lesson throughout the upper elementary grades. The study of physics in each of the three upper grades has a certain flavor and concentrates on one aspect of the thought process.

Sixth graders mostly need to draw *conclusions* from their sense perceptions. The teaching of physics in the sixth grade is, therefore, almost entirely based on *observation*. Seventh graders most need to be shown how to form sound *judgments*. The teaching of physics in this grade therefore incorporates opportunities for *quantification and measurement*. Because eighth graders are ready to develop clear *concepts*, they need to be challenged to derive concepts and laws from their *experiences* and from the world around them.

I think that Rudolf Steiner suggested that we cover all of the subjects each year so that concepts could grow and develop. What does this look like in practice? If we look at the study of heat, for example, we see that the curriculum for the sixth grade focuses on the students' experiences of warmth and cold. The students see the effects of heat on various substances and learn how heat travels. In the seventh grade the study of heat focuses on the measurement of heat. In the eighth grade, we can study the use of heat by examining heating systems. A hot water heating system, for instance, embodies and applies all the concepts and principles that the students have learned up to this point. Were this to be studied in the sixth grade, it would not have the richness or the resonance of experience that the eighth graders have gained over the course of the previous years.

If such an approach is to be successful, teachers need to develop a view of the entire physics curriculum for the three years. They must choose their subjects carefully and practice economy in their teaching.

If you spend one week on the study of heat in the sixth grade, for example, you would probably only spend two or three days in the seventh, and one or two in the eighth grade. This total of two weeks, however, allows much more material to be covered—and to be covered in greater depth—than two weeks spent on heat in the sixth grade.

There is another compelling reason to study all the subjects every year, adding new topics each year: the forces of sound, light, heat, magnetism, electricity, and mechanics have many similarities. These similarities can be most clearly elucidated by experiencing all of the subjects each year. For instance, if students study the way that sound, light, and heat travel, they can see the similarities: all of these forces can be obstructed, deflected, reflected, or absorbed. From these sorts of observations students begin to arrive at clearer, more comprehensive ideas of the nature of the forces themselves.

Finally, it is important that the students experience the evolution and transformation of a subject over a period of years. This is one rationale for covering some of the same material in the seventh grade that was covered in the sixth. In the mirror of different approaches to the material, the students are able to see a reflection of how they have grown and changed. In this same mirror the students may see a reflection of how their class teacher has been able to understand their changing perspective and needs and has worked to address and fulfill them.

From my experience, I have concluded that it is not only possible, but advisable, to cover the subjects that Rudolf Steiner suggested each year. If we follow Rudolf Steiner's advice stated above, if we keep a view of the whole, if we choose the subjects carefully, and vary the approach each year, we should be able to help the students arrive at a meaningful synthesis of all that they have learned by the end of eighth grade.

The major caveat to this conclusion is when physics is being taught by someone other than the class teacher. The advantage of having a specialist teach this subject is that the person typically tends to have a much more thorough background and understanding of the subject; however, there are two major disadvantages. The specialist is not likely to have a sense of the seeds of the concepts that have already been planted by the class teacher, nor will she know the students nearly as well as the class teacher.

Having taught many blocks as a guest teacher in the seventh and eighth grades, I have found it better to focus only on the two new subjects rather than to try to include the previous subjects. In seventh grade this meant focusing on electromagnetism and mechanics and in eighth grade on fluid mechanics and aeromechanics. This approach allowed me to develop the subjects more fully and to give the students a sense that they have learned a great deal.

In the sections that follow, I have developed the curriculum through a syllabus and a schedule that I found to be ideal for building the students' conceptual understanding of the subjects.

The Task of the Teacher

Teachers who are preparing to teach physics in the upper elementary grades often feel intimidated by the material, overwhelmed by the amount of preparation, and anxious or afraid of the challenges ahead. In the face of such feelings, it may help to remember that the most important part of our teaching—more important than the curriculum, more important than our methods and materials—is the way we work with and relate to matter. Not only do our students watch us handle the equipment as we demonstrate a process, they also internalize our approach and are influenced by our attitude.

Teaching physics can be a wonderful, delightful, and rewarding process—interesting, exciting, and fun. Teachers who overcome their trepidation and prepare properly will begin to find that they feel at home in a new world, that they are able to reveal the mysteries of this world, and that they are increasingly able to control the forces of nature. With this increasing sense of comfort and competence come two types of temptations.

The first temptation is to become like a magician. As we become more and more comfortable in the sciences, we can begin to feel that we have a magic wand with which to reveal the mysteries of the universe. See how we amaze the students! Look how we mesmerize and enchant them! How lovely to bask in the glow of their admiration, their wonder, their awe!

How easy it is to succumb to this temptation, for all of us yearn for our lessons to take wing, for our words and deeds to illuminate and inspire. Yet if we do, we can turn our students into mystics and dreamers with their heads in the clouds. Our task is to help our students connect to the earth, not become divorced from it.

The second temptation is to become like a power-hungry mad scientist. As we become more proficient in the sciences, as we begin to know and anticipate the processes and outcomes of our experiments, we can begin to feel that we have the means of controlling the forces of nature. Look how nature does our bidding! See how we can manipulate and regulate the phenomena! How satisfying to feel the students' increased respect for our power!

How easy it is to succumb to this temptation, for all of us strive to be competent, to master our task and our material, to do things well. Yet if we do so, we begin to turn our students into materialists with their feet rooted in the earth, their gaze focused downward. Our task is to help students find the spirit in matter, not to become materialists.

We are neither magicians nor mad scientists—we are teachers, teachers who are being called on to serve as priests of Nature, who work in Nature's service and on her behalf. We do not need to pull back the veil of the temple; we can let Nature reveal her mysteries in her own way. We do not need to manipulate the phenomena on the altar of the demonstration table; we can simply perform our service and let the phenomena speak for themselves.

When students observe a demonstration, they should observe the phenomena, not the teacher. We need to learn to stand aside, literally and figuratively, so that Nature may speak directly to the students' hearts and minds.

There is the middle way between these two temptations—the way of humility. By seeking and nurturing our true humanity, we will be helped in our endeavor to teach our students about the world and to help them learn to relate to matter in a moral way.

To educate youth
Means to foster and tend
In matter the spirit,
In today the tomorrow,
In earthly life,
the spirit's existence.

– RUDOLF STEINER

The Study of Physics in the Sixth Grade

The study of physics in the sixth grade introduces students to new ways of seeing the world. Students are going to become newly aware of phenomena that they have experienced before. By becoming aware of these phenomena in new ways, they will begin to become more objective and exacting in their thinking.

Because we are calling on a new faculty in the students—the ability to objectify experience—we have to proceed slowly and simply so that students can develop confidence in their perceptions and their ability to think.

The fundamental questions for the study of physics in the sixth grade are WHAT and WHEN. What did you experience? What did we do to bring this experience about? When else have you experienced this? When can you expect it to happen again? These questions have to do with the most basic levels of a phenomenon and with the process and the first stages of forming judgments. Sixth graders may begin to have a sense for the concepts underlying the phenomena they are studying, but at this stage, the concepts are not as important for them as the variety, the beauty, and the power of their experiences.

In the sixth grade, three main aspects of the physical forces are considered: their nature, their sources, and how they travel. We want students to realize how each force arises in the world and what we must do to bring it into being. We want them to see how these forces travel and to note the qualities of their transmission. Only if this foundation is laid will students be able to understand the interactions of the physical forces when they study physics in the seventh grade and the applications of physical forces in the eighth grade.

As adults, we tend not to experience the world as directly and vividly as we once did. We tend to forget how much we view and understand the world through the concepts we have developed over the course of many years. With our well-defined concepts, it is easy to present material to students in a fixed, finished, conventional form. We have to resist this tendency and help students discover and develop their own concepts, concepts that can grow.

The teaching of physics in the sixth grade is primarily about experiences rather than concepts. If we allow students to experience the phenomena and help them bring their experiences to consciousness, if we help them reflect upon their experiences and begin to become more objective about them, we will have done a great deal. There is time enough in the seventh and eighth grades to deal with the more "scientific" aspects of physics. For now, the introduction to physics should be interesting; it should be exciting; and most of all, it should be fun.

Introduction to the Study of Sound

The study of sound in the sixth grade is the students' introduction to physics. As Rudolf Steiner said in his *Curriculum Lectures*: "We start with the teaching of physics in the sixth class in such a way that it is linked with what the child has acquired through the teaching of music. We start our teaching of physics by letting acoustics grow out of the musical."

In preparation, students can be challenged to listen to music more consciously. As they do, they start to notice the different kinds of sounds, their attributes, and their effects. Teachers can ask questions: What kinds of sounds are there? How do they originate? Which instruments create which kinds of sounds? How are different sounds created on an instrument? These questions lead students from the experience of music to the study of sound as a physical force.

We begin with the origins of sounds and what has to occur for sounds to arise. Gradually students will begin to recognize the connections between sound and vibration, pitch and frequency, volume and amplitude. Then we can examine how sound travels. The fact that it does travel may be a new realization for some students. As they experiment to see how sound travels through various media, the students' experience of sound is broadened and deepened.

The week spent on sound can end with a mystery: the Chladni plate. (See *Demonstration 19. The Chladni Plate.*) It is sufficient for sixth-grade students simply to witness this mystery; they need not understand it. The changing geometric forms engendered by sounds are a revelation that should not be deflated by intellectual explanations. Understanding will eventually follow experience; we don't need to rush it. If we allow students to simply dwell on the profundity and the beauty of experiencing the phenomena, we prepare them to consider other physical forces with a heightened level of awareness and interest.

TOPICS

The week allotted to the study of sound includes the following topics:

- Introduction: the larynx
- The sources of sound
- Pitch and volume
- The propagation of sound
- The formative aspects of sound

CHARACTERISTICS OF SOUND

In working with these topics, the following characteristics will be revealed:

- Sounds have different sources.
- Objects can be identified by the sounds they produce.
- Sounds can be created by blowing, plucking, bowing, striking, or rubbing.
- The pitch of sounds can be altered by changing some aspect of the sound-producing body.
- The volume of sound depends on the intensity of production.
- Sound travels differently through various media.
- The effects of sound can be made visible.
- The timbre of a sound is responsible for its unique qualities.

These characteristics can be revealed, but they cannot be explored fully in the course of one week. In sixth grade we only need to introduce the students to acoustics, not engage in an exhaustive study. This unit should help the students become more aware of the world of sound; they will have the rest of their lives to deepen their experience and their understanding.

DAY	DEMONSTRATIONS AND ACTIVITIES	CHARACTERISTIC
1	■ The Larynx ■ Listening ■ Mystery Sounds	■ Introduction to sound. ■ Sounds have different sources. ■ Objects can be identified by the sounds they produce.
2	■ Creating Sounds by Various Means	■ Sounds can be created by blowing, plucking, bowing, striking, or rubbing.
3	■ Pitch and Volume of Different Types of Sound	■ The pitch of sounds can be altered by changing some aspect of the sound-producing body. ■ The volume of sounds depends on the intensity of production.
4	■ How Sound Travels through Air and Water	■ Sounds travel differently through various media.
5	■ How Sounds Travel through Solids	■ Sounds travel differently through various media.
6	■ The Chladni Plate	■ The effects of sound can be made visible.

Additional or alternative activities are included at the end of this chapter.

DAY 1

INTRODUCTION: THE LARYNX

Rudolf Steiner suggested that we begin the study of sound by letting acoustics grow out of the musical. According to Steiner, you can "bring acoustics into relation with musical science when you describe the human larynx physically and physiologically."

When studying the larynx, teachers should describe how we produce sounds and how we regulate pitch and volume. This sets the foundation for the topics that will be developed throughout the course of the week. After the larynx has been described, we can compare human sounds with other types of sounds.

ACTIVITY

1. Listening

CHARACTERISTIC: Sounds have different sources.

Sounds can be grouped according to their origins or means of production. The broad areas of classification divide sounds into those produced by natural elements (wind, rain, thunder, etc.); those produced by living beings (animals, humans); and those produced by instruments, machinery, and technology.

In preparation for students' first experiences in acoustics, discuss what can be learned about the world through sound. Sounds can reveal to us the power of the elements; the emotions of animals; the thoughts, feelings, and intentions of human beings; the qualities of materials and the uses to which they are put; and the nature of activities and processes.

Conscious listening is a wonderful activity to introduce students to the world of sound. Although the sounds they listen to will probably not be new, when they do this activity, they will become more conscious and aware of their perceptions.

MATERIALS

1 pencil per student

1 piece of paper per student

PROCEDURE

Students should sit quietly in a comfortable position and listen for three to five minutes. As they hear sounds, they should write down the sounds and what made them. When they are done, students should share what they have heard.

COMMENT

This is a good homework activity, which can be done on the weekend preceding the study of acoustics or on the first day of the main lesson block. After students share their lists of sounds with the class, the sounds should be grouped according to their origins. The list of sounds can be written in a poetic form.

ACTIVITY

2. Producing Various Sounds

CHARACTERISTIC: Sounds have different sources.

MATERIALS

1 noise-making object per student such as a whistle, bell, chime, drum, etc.

PROCEDURE

At a given signal, all the students should make a sound with their objects. At another signal, they should stop. Then each student should be given a chance to make a sound individually so that the different sounds can be heard distinctly.

Students should group themselves with others who are making similar sounds. Then each group should make sounds together.

As the class now listens to the different groups of students making their sounds, they will notice the similarities among the objects that produce the sounds and among the methods of sound production.

During the review the next day, the differences and similarities among the sounds should be identified. Ask students to group the sounds into musical and non-musical categories. Ask students to think about what makes a sound "musical."

Students can also group sounds in terms of their sources or methods of production. For example, a sound may be created by blowing into or over an object, by rubbing two objects together, or by striking things together.

COMMENT

Students will note that when many sounds are created at once, it is difficult to distinguish individual sounds. Those sounds that stand out from the mass of sounds can usually be identified because of the extremity of their pitch or volume. When sounds are made individually, many more of their salient characteristics can be identified.

Students will probably agree that what makes a sound musical or non-musical is the distinctness of the tones or pitch. Sounds with distinct pitches tend to be considered musical, while those whose pitches cannot be distinguished tend to be regarded as noise. Purely percussive sounds can be used to create or enhance music when they reinforce the rhythm or beat of a musical composition.

ACTIVITY

3. Mystery Sounds

CHARACTERISTIC: Objects can be identified by the sounds they produce.

Sounds reveal something about the nature of the objects that created them. This is one of the most important characteristics that should be developed in the study of sound. In order for students to identify the origins and aspects of the sounds around them, their listening faculties must be sharpened.

This is a very good activity to begin the study of sound, but refrain from including too many types of sounds because the students will begin to lose their concentration or interest.

MATERIALS

- 10–12 objects that make different kinds of sounds
- 1 pencil per student
- 1 piece of paper per student

PROCEDURE

Either set up a screen behind which you can work or have the students sit facing away from you so they cannot see you. Make a sound with one of the objects. The students should try to identify the object they think created the sound and write down its name. Repeat with the other objects. When you are finished, have students share their guesses and then reveal the objects.

COMMENT

In the review the following day, ask students to characterize the sounds and to express their different qualities. Since so many words for sounds are onomatopoetic, the class can develop a vocabulary for the world of sound. These introductory activities should help the students realize that the nature of an object can be revealed by the sounds it makes.

DAY 2

DEMONSTRATION

4. Sounds Can Be Created by Blowing

CHARACTERISTIC: Musical sounds can be created by blowing, plucking, bowing, striking, or rubbing.

In previous activities, students were introduced to different sources of sound, and they were asked to discriminate between musical and non-musical sounds. Now they will be asked to make more exact observations.

It is hard to resist presenting the students with too many sound experiences, but if they experience too many different demonstrations, they will end up remembering little.

MATERIALS

several objects that can be used to create blown sounds such as bottles, whistles, pen caps, etc.

several types of wind instruments

PROCEDURE

Make a sound with each of the objects and with each of the musical instruments.

DEMONSTRATION

5. Sounds Can Be Created by Plucking, Bowing, or Striking a String

CHARACTERISTIC: Musical sounds can be created by blowing, plucking, bowing, striking, or rubbing.

After showing how sounds can be made by blowing, we can focus on sounds created by taut strings.

MATERIALS

several different types of stringed instruments

PROCEDURE

Pluck and bow the strings of various instruments.

COMMENT

Students should note the similarities and differences between the sounds created by plucking and bowing. Students might be shown the hair of a violin or cello bow under a microscope to see the little barbs that repeatedly pluck the string when a bow is drawn over the strings. This will help them recognize that bowing is just repetitive plucking.

DEMONSTRATION

6. Sounds Can Be Created by Striking

CHARACTERISTIC: Musical sounds can be created by blowing, plucking, bowing, striking, or rubbing.

MATERIALS

several objects that make sounds when struck, such as a block of wood, a pipe, or a bell

several types of percussion instruments

PROCEDURE

Make a series of sounds with each of the objects and instruments.

COMMENT

Percussive sounds provide vivid examples of how the quality of a sound is determined by the material or object used to make the sound.

DEMONSTRATION

7. The Singing Glass

CHARACTERISTIC: Musical sounds can be created by blowing, plucking, bowing, striking, or rubbing.

As a culmination of the day's demonstrations, you can make a wine glass sing or use a singing bowl to produce ethereal notes. The students will easily recognize the connection between the purity of the materials and the qualities of the tones they produce.

PROCEDURE

To make the wine glass produce a beautiful sound, partially fill it with water and move a finger around the rim smoothly and firmly. (Wetting your finger slightly with vinegar can help create the necessary friction.)

To make the singing bowl sing, gently strike the edge with the mallet. Then rub the mallet lightly but firmly around the outside rim of the bowl.

MATERIALS

- several goblet-shaped wine glasses
- water
- vinegar
- 1 singing bowl and mallet

COMMENT

The glass harmonica, developed by Benjamin Franklin, uses the same principle as the rubbed wine glasses. Students can listen to Tchaikovsky's "Dance of the Sugar Plum Fairy," which features this instrument.

DAY 3

DEMONSTRATION

8. How Pitch Is Created in Blown Sounds, Part 1

CHARACTERISTIC: The pitch of sounds can be altered by changing an aspect of the sound-producing body.

Some students have difficulty producing a sound by blowing in a bottle, so it is better for the teacher to blow into the bottles for this demonstration. The bottles can be left out for the rest of the week and students can try to make clear sounds by blowing into the bottles themselves.

MATERIALS

a set of glass or plastic bottles of various sizes

PROCEDURE

Have the students suggest how to arrange the bottles of various sizes in order of ascending or descending pitch. Blow into each bottle to ascertain its pitch and make any adjustments in the order.

COMMENT

By using bottles of vastly different sizes, students will gain a more vivid visual impression of the relationship of volume to pitch. Be aware that there may be some surprises in the pitch of the bottles according to their size. For instance, a large bottle may make a high-pitched tone. These variations are usually due to the sounding of the overtone—a note that sounds along with the primary or fundamental note—or to a change in the length of the column of air.

DEMONSTRATION

9. How Pitch Is Created in Blown Sounds, Part 2

CHARACTERISTIC: The pitch of sounds can be altered by changing an aspect of the sound-producing body.

Having shown that bottles of different sizes create different pitches, we can refine the students' observations by working with bottles of the same size. Here the only variable that will change is the amount of air in the bottle. This demonstration can easily follow the previous one without interruption. In this demonstration, the amount of air changes as we add water to the bottle.

MATERIALS

a set of bottles of the same shape and size

water

PROCEDURE

Blow into several of the empty bottles to show that their pitches are the same. Fill one of the bottles a quarter full of water and blow into it. Fill the second bottle with more water than the first and blow into it. Fill all the bottles with varying amounts of water and have the students predict how to arrange them in order of their pitches.

COMMENT

In this demonstration, some students are likely to misunderstand what affects the pitch. They may need help to recognize that the amount of air in the bottles determines the pitch, not the amount of water. Adding water simply decreases the amount of air in the bottle.

Once students understand the connection between the pitch and the height of the column of air, they may begin to understand how pitches on wind instruments are created. Covering or uncovering holes on a recorder, for instance, changes the length of the air column and affects the pitch. (Pitches on some wind instruments may also be changed by blowing harder. See the grade 7 unit on sound, specifically on the subject of overtones.)

DEMONSTRATION

10. How Pitch Is Created in Plucked Sounds

CHARACTERISTIC: The pitch of sounds can be altered by changing an aspect of the sound-producing body.

PROCEDURE

Pluck the different strings on an instrument to show the relationship between the thickness of the string and its pitch. Tighten the string to show the relationship between tension and pitch. Shorten the string by placing a finger on the fingerboard to show the relationship between length and pitch.

MATERIALS

1 or more string instruments

COMMENT

Some students may note that the duration of a string's vibration also depends on its thickness, tension, and length. In the review the next day, ask students to reflect on the differences among instruments and the roles they play in an orchestra or chamber group. For instance, high-pitched instrument, such as violins, are typically used to play the melody because they have shorter, thinner strings that give players more mobility. Instruments whose strings vibrate longer, such as cellos or basses, are used to create the harmony.

DEMONSTRATION

11. How Pitch Is Created in Percussive Sounds

CHARACTERISTIC: The pitch of sounds can be altered by changing an aspect of the sound-producing body.

Most sounds created by striking have a certain degree of pitch, but the pitches are often not clear because they are too high, too low, or too complex. Some percussive musical instruments, such as the xylophone, have clear pitches, however, and those are the ones we want for this demonstration.

MATERIALS

1 glockenspiel or xylophone

a set of glass or plastic bottles of the same shape and size

water

PROCEDURE

If it is possible to obtain individual bars of a glockenspiel, arrange them in a random order and play the various notes. Then ask one of the students to arrange the bars in order of ascending or descending pitch.

Students can also fill a series of equal-sized glass bottles with varying amounts of water so that a scale is produced by blowing in the bottles, as they did in *Demonstration 9: How Pitch Is Created in Blown Sound, Part 2*. Next, they should tap the bottles. They will notice that when the bottles are tapped, those that made the highest blown sounds—those with the most water and the least air—will have the lowest tapped sounds, and the scale will be heard in reverse.

DEMONSTRATION

12. How Pitch Is Created in Sounds Created by Friction

CHARACTERISTIC: The pitch of sounds can be altered by changing an aspect of the sound-producing body.

The principle of pitch created by friction can be illustrated well by wine glasses of different sizes, or by wine glasses that are the same size but have different amounts of water in them.

PROCEDURE

Make the large glass sing by rubbing a finger moistened in vinegar around the top. Repeat with the small glass.

Create a note by rubbing one of the wine glasses, then pour in some water and create another note. Add more water and create a third note.

MATERIALS

2 wine glasses,
1 larger and
1 smaller

water

vinegar

COMMENT

Students may have trouble recognizing that the variations in pitch are due to the total mass of the glass and water. As the mass increases by the addition of water, the pitch is lowered. This is different than for blown sounds, where more water in the bottle creates a shorter air column, resulting in a higher pitch.

Connecting Volume and Intensity of Sound

In all of the previous demonstrations, students may have recognized the relationship between volume and the intensity with which the sound is produced. When the demonstrations are reviewed, this phenomenon can be described and its implications discussed. The volume of sound depends on the force with which the sound is produced.

DAY 4

ACTIVITY

13. How Sound Travels Through Different Media

CHARACTERISTIC: Sound travels differently through various media.

Having explored the origins of sound, the students can begin to explore the way sound travels through various media. This is a much more abstract aspect of acoustics, for students are not apt to think about the fact that sound travels.

Although sound does not travel as well through the air as it does through solids or liquids, enclosing the air space improves the propagation of the sound tremendously. This is the principle utilized by speaking tubes and stethoscopes.

MATERIALS

1 garden hose, at least 25 feet long

2 large funnels (optional)

masking tape or duct tape

PROCEDURE

Wrap tape around the small end of each funnel so it fits snugly into the end of the hose. Then tape the funnels to the ends of the hose. Have two students take the ends of the hose and stretch the hose around a corner or out the window so that the students are as far apart as possible. Have one student speak to the other through the hose. The speaker can give the listener instructions to see whether his or her words are heard and understood.

ACTIVITY

14. Sounds at a Distance

CHARACTERISTIC: Sound travels differently through various media.

An effective way to show that sound actually has to travel from its source to the listener is to experience the time gap between seeing a sound being produced at a distance and hearing it. This activity should be done outside.

PROCEDURE

A student should practice making a loud clacking sound by hitting the boards together, making sure that his or her fingers are not in danger of being caught between the boards. Position the student at a spot close enough to be seen clearly—about 200 yards—but far enough away that there is a delay between seeing the boards hit and hearing the sound they make.

MATERIALS

2 pieces of 2″x4″ lumber, each about 1 foot long

ACTIVITY

15. Bath Time

CHARACTERISTIC: Sound travels differently through various media.

Students will be surprised to learn that most sounds can be transmitted through liquids. This activity is a good homework assignment after learning about how sounds travel through the air. When this activity is reviewed, students can also discuss how marine mammals communicate.

MATERIALS
a bathtub filled with water

PROCEDURE
Students should take a bath and experience what can be heard above and below the water. They can listen to a sound with their ears above water and then experience the same sound with their ears submerged. They can listen to a sound created under water and then lift their heads and experience the underwater sound with their ears above water. They might notice that when their ears are submerged, they can hear many sounds that are not audible when their ears are above water.

DAY 5

ACTIVITY

16. How Sound Travels Through Solids

CHARACTERISTIC: Sound travels differently through various media.

Having dealt with how sound travels through the air, students can experience how sound travels through solid substances. Students may initially think that solids do not conduct sound, but in this activity they will hear how sounds can travel through solids better than through the air.

MATERIALS
classroom desks
1 yardstick

PROCEDURE
Make a very soft sound with the yardstick by raking it across the floor. Have the students listen. Make the sound again as they listen pressing one of their ears against the surface of their desks and covering up the other ear.

COMMENT
The students will be able to hear the sounds you make much more loudly through their desks than through the air. If the room is still, they may even be able to hear sounds coming from other rooms.

ACTIVITY

17. Cutlery String Chimes

CHARACTERISTIC: Sound travels differently through various media.

This is one of the most surprising and effective demonstrations in the study of sound. Until you have tried it, you cannot imagine the richness and depth of the sounds that can be made by simple pieces of cutlery.

MATERIALS

several pieces of string about 2 feet long

several metal tablespoons and teaspoons

several metal table forks and knives

metal rulers

PROCEDURE

Students should pass a piece of string through the hole of a metal ruler or tie it around the handle of a knife, fork, or spoon. Then they should wrap each end of the string around the ends of their index fingers. Allowing the metal object to hit against another object, they should listen to the sound.

Next, students should insert their stringed index fingers in their ears, bending over slightly so that the strings do not touch anything else. They should allow the metal objects to hit their desks and listen to the sounds of the clanging objects as they are transmitted through the strings and their fingers into their ears.

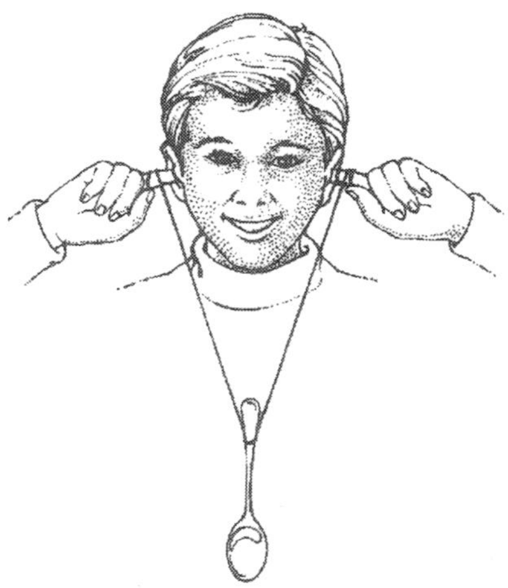

COMMENT

Students will notice not only that the sound is clearer and deeper when it is transmitted through the string and their fingers but also that they can hear several pitches besides the fundamental note.

ACTIVITY

18. Making a String Telephone

CHARACTERISTIC: Sounds travel differently through various media.

Many students will have tried this activity before, but now they can recognize the acoustical principles underlying the phenomenon.

MATERIALS

large plastic cups
paper clips
finishing nails
string, monofilament line, or magnet wire in lengths of 20–25 feet

PROCEDURE

Make a small hole in the bottom of each cup using a nail. Insert one end of the string through the hole and tie a knot large enough to stop the string from slipping through the hole when the string is taut. If using wire, the end of the wire can be wrapped around paper clips. Now attach a cup to the other end of the string or wire in the same way.

Stretch the string or wire taut. Involve two students, each holding a cup. One person whispers into his cup while the other holds his cup to his ear and listens.

This type of telephone works best when the line remains taut and doesn't brush against the rim of the cup.

COMMENT

Students may wish to compare how sound travels through different strings and wires to investigate which materials transmit sound the best. To store the telephone, carefully wind the string around one cup and insert that cup inside the other.

It is also possible to make a party line by attaching two or more strings and cups to the end of the original line.

DAY 6

DEMONSTRATION

19. The Chladni Plate

CHARACTERISTIC: The effects of sound can be made visible.

A Chladni plate is a metal plate on a stand that can be clamped to a table. Chladni plates may be made in a number of different shapes out of different metals, and each of the different shapes and metals reveals a different set of Chladni figures.

In my experience, this demonstration is extremely effective, and students are impressed and awed by the creation of the patterns appearing on the plate. However, it is not easy for them to understand what causes the patterns. I suggest that the Chladni plate be the culminating demonstration on the sixth day after the final review of the study of sound. This demonstration leaves students with the aesthetic impression of the patterns without requiring an explanation of their causes.

These Chladni plate demonstrations lay a foundation for the study of overtones and frequency, and for the understanding of the connections between sound and vibration. In the eighth grade, the Chladni plate can be examined again, and students can be asked to explain why the patterns form.

PROCEDURE

A Chladni plate can be obtained from a science supply company. Make sure that the Chladni plate is firmly attached to its stand. The plate needs to be exactly parallel to the ground, otherwise the salt will shift to one side. Place the tray under the Chladni plate to catch the salt.

To make a sound on the Chladni plate, draw the bow across the edge of the plate, pressing fairly hard and holding the bow perpendicular to the plate. If the bow is drawn along one spot on the edge of the plate, it will create a tone. The pitch of the tone varies according

MATERIALS

1 Chladni plate
1 violin or cello bow
rosin
salt in a salt shaker
1 tray

to the position along the edge where it is drawn. Dampening, pinching, or bracing the plate from underneath with a finger helps create different pitches.

To reveal the Chladni figures, sprinkle salt on the plate and then create a tone with the bow. As the sound is produced, the plate vibrates and the salt forms interesting and beautiful patterns. Different patterns will emerge as the tones or blends of tones change. As you gain experience with the Chladni plate, you will be able to move from simpler to more complex patterns and to predict what patterns will be generated by which tones.

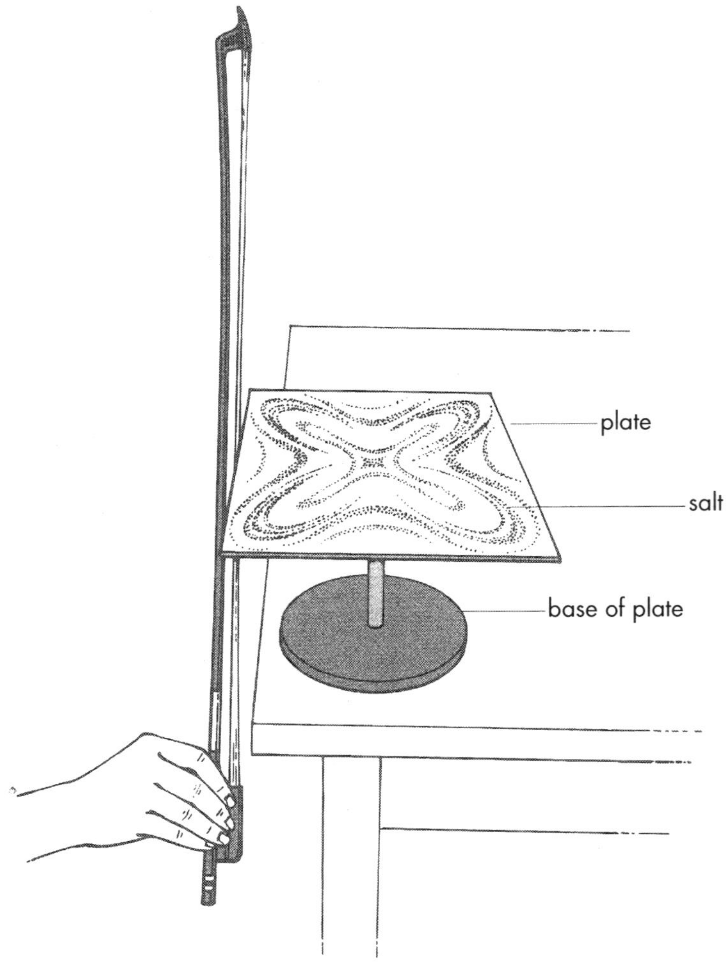

COMMENT

You need to practice working with the Chladni plate in order to be confident about producing the tones. Marking the spots where the bow is drawn can help you remember where to place the bow to create the various tones.

ADDITIONAL & ALTERNATIVE ACTIVITIES

Some of the following activities can replace those suggested in the previous section. Others can be done in conjunction with the review or assigned as homework or extra projects.

ACTIVITY

20. Making a Bull Roarer

CHARACTERISTIC: Sounds can be created by blowing, plucking, bowing, striking, or rubbing.

You might wish to develop the theme of how sounds are produced by having the students make a bull roarer, a popular folk toy found in many cultures. Students tend to underestimate the distance covered by the string and ruler, so be sure students stand well back when observing the bull roarer.

PROCEDURE

Tie a piece of string about 2 feet long through the hole in the end of the ruler. Hold one end of the string and slowly whirl the ruler in a circle perpendicular to the floor. At a certain point the ruler will begin to twist and whirl, creating a roaring sound. The twirling speed can then be increased or decreased, changing the pitch of the bull roarer.

MATERIALS

- 1 metal or wooden ruler with a hole near one end
- 1 piece of sturdy string, about 2 feet long

DEMONSTRATION

21. The Slide Whistle

CHARACTERISTIC: Sounds can be created by blowing, plucking, bowing, striking, or rubbing.

The slide whistle is a fun way to show how changes in pitch are determined by the length of the air column, but students may need help to recognize how it works.

MATERIALS
1 slide whistle

PROCEDURE

Make a series of sounds demonstrating that the pitch becomes higher or lower as the slide goes in or out.

ACTIVITY

22. A Jug Band

CHARACTERISTIC: The pitch of sounds can be altered by changing some aspect of the sound-producing body.

This activity can be done in the days or weeks following the demonstrations with blown sounds. It is a good activity to be done at a school assembly because it utilizes the principles learned in the study of acoustics. Not all students are able to easily produce a note by blowing into a bottle, though, so some practice may be needed.

MATERIALS
bottles of different shapes and sizes
water

PROCEDURE

Tune the bottles to the notes of a well-known piece of music, such as "Twinkle, Twinkle, Little Star," "Hot Cross Buns," or "Row, Row, Row Your Boat." Give the bottles to the students and have them practice blowing in them until they can produce notes at will. Write the music on the board or conduct the piece by pointing to the student who should make the next note. After several practices, the students can play without a conductor.

ACTIVITY

23. A Glass Harmonica Ensemble

CHARACTERISTIC: The pitch of sounds can be altered by changing some aspect of the sound-producing body.

PROCEDURE

Tune a set of six to eight wine glasses by adding water so that they play the notes of a simple piece of music. Let the students practice producing sounds on the glasses by rubbing their fingers along the rims. Then teach them to play the piece of music.

MATERIALS

6–8 wine glasses of equal size
water

ACTIVITY

24. Visible Sound

CHARACTERISTIC: The effects of sound can be made visible.

Sound can be made "visible" with a candle flame. This activity can be assigned as an extra project to some of the students who can then report on it or show it to their classmates.

PROCEDURE

Light the candle and set it in a place with no drafts. Hold the open end of the container near the candle flame and tap the other end. *(**CAUTION:** Be careful to not place the container so close to the flame that it gets singed.)* As a sound is produced, the moving air in the container disturbs the candle flame.

MATERIALS

1 round cardboard oatmeal box or 1 round plastic container that is closed at one end
1 candle and candle holder
matches

COMMENT

The students will observe the candle flame move each time the container is tapped. While it is not altogether obvious that the sound is traveling through the air, it is nevertheless clear that the air has been set in motion by the tapping. In this context, one might consider the effects of sounds traveling through air, e.g., by thunderstorms, sonic booms, etc.

Timbre

Timbre is the most subtle aspect of sound that might be considered in the sixth grade. To work with timbre in a meaningful way, students will have to listen carefully and to express their perceptions clearly and accurately.

When we hear a note played on an instrument, it is easy to identify on which instrument it was created because each instrument has a distinctive tone quality or timbre. The following activities can be turned into a listening and guessing game.

ACTIVITY

25. Recognizing Tone Quality, Part 1

CHARACTERISTIC: The timbre of a sound is responsible for its unique qualities.

MATERIALS

1 set of instruments that have the same musical range, e.g., violin, flute, recorder, and oboe.

pencils

paper

PROCEDURE

Have the students who are listening turn their chairs around so that they cannot see which instrument is being played. Several students should then play a note of the same pitch but on different instruments. Have the listening students write down their guesses of which instrument made the sound.

COMMENT

Students will recognize that although the different sounds had the same pitch, each one sounded very different from the next.

ACTIVITY

26. Recognizing Tone Quality, Part 2

CHARACTERISTIC: The timbre of a sound is responsible for its unique qualities.

Even with identical instruments, small variations in the production of the sound will result in differences in tone quality.

PROCEDURE

Have a few students play the same series of notes on their instruments, and have the other students see if they can recognize the differences between the players.

MATERIALS

1 set of identical instruments, e.g., recorders.

Notes

The Study of Light in the Sixth Grade

After studying sound for a week, we move on to the study of light and color. Hearing is a more unconscious process than is seeing, so students will engage in the study of light in very different way. It is much easier for students to become aware of their visual impressions than their impressions of sound.

It is difficult for us to experience total quiet, for it is almost impossible to shut out the noises that surround us. Even if we cover or plug our ears, sounds can be heard and felt. However, every time we close our eyes the world becomes dark. We will build on that common experience of darkness to reveal important qualities about light. If the students are properly prepared, this week's experiences can engender a deep inner response.

Have students sit in the darkness for a while, then light a small flame. When students see the smallest light dispel the darkness, they will experience one of the parables of nature about which Rudolf Steiner spoke, how great truths can be perceived in physical phenomena. The study of physics is full of such parables, if students can awaken to them. We shouldn't present such parables openly; instead we should strive to create the situations where the students can experience them. Later we can lead students to reflect on their experience so that inner recognition can grow.

After experiencing darkness, students will be introduced to some of the properties of light: how it can only be seen at its source or by what it illuminates; how it travels in straight lines in all directions from its source; and how some materials allow light to pass through, while others obstruct light.

The students have experienced many of these phenomena in their lives, but they were probably not fully conscious of them. The process of awakening to this new world is a joyful one and should not be cluttered with too many facts and concepts.

Out of the interplay of light and darkness comes color, and much of the week will be devoted to the experience and the study of color. This is a familiar realm, for students have worked with color for many years, yet it is unlikely that they are aware of the laws of color. In first grade, when students compared how yellow and blue or yellow and green looked together, their teacher may have stated that the combination of yellow and blue is very pleasing. Such a judgment was simply accepted.

Now, in sixth grade, as they learn the laws of primary, secondary, and complementary colors, students will begin to understand the basis for such judgments. The week of studying color ends with afterimages and colored shadows—beautiful and unexpected phenomena that will awaken wonder and inspire interest in the students.

TOPICS

The study of light in the sixth grade treats three major topics:

- The nature of light and darkness
- Properties of light
- The nature and properties of color

CHARACTERISTICS OF LIGHT

In working with these topics, these characteristics will be revealed:

- Light can only be seen at its source or by what it illuminates.
- Unless obstructed, light travels in straight lines in all directions from its source.
- The intensity of light decreases with distance.
- Materials are variously permeable to light: some reflect light; some allow light to pass through; some obstruct light.
- Colors arise from the interplay of light and darkness.
- Colors can originate when light passes through a prism.
- Colors may be grouped into primary and secondary colors.
- Primary colors can be mixed to form secondary colors.
- We can experience afterimages and complementary colors.
- Our perception of the color of an object depends on the color of the light illuminating it.
- Shadows are created by an obstruction of light, and colored shadows can be created by a mixture of colored lights
- The darkness of a shadow is determined by the intensity of the light.
- The shape of a shadow bears a direct relationship to the shape of the object casting the shadow.

You could spend a month—or many months—studying light because there is so much to discover and experience, but in order to give the other subjects their due, I recommend spending only a week on this subject. Fortunately, the students will continue to study light in the seventh and eighth grades, and as they return to this subject with greater maturity, they will be able to understand much more than they did in the sixth grade.

The topics listed above can be introduced by the following demonstrations and activities. Additional or alternative activities are included at the end of this chapter.

DAY	DEMONSTRATIONS AND ACTIVITIES	CHARACTERISTICS
1	■ Total Darkness	■ Introduction to the nature of light and darkness.
2	■ The Unrevealed Revealer ■ The Black Box ■ The Straight and Narrow Path ■ Transparent, Translucent, & Opaque	■ Light can only be seen at its source or by what it illuminates. ■ Unless obstructed, light travels in straight lines in all directions from its source. ■ Materials are variously permeable to light: some reflect light; some allow light to pass through; some obstruct light.
3	■ The Generation of Color ■ Creating Colors with a Water Prism ■ Working with Prisms	■ Colors arise from the interplay of light and darkness. ■ Colors can originate when light passes through a prism.
4	■ Mixing Liquid Pigments ■ Mixing Light	■ Colors may be grouped into primary and secondary colors. ■ Primary colors can be mixed to form secondary colors.
5	■ Afterimages ■ Afterimages from Afterimages	■ We can experience afterimages and complementary colors.
6	■ Colored Shadows	■ Colored shadows can be created by the mixture of colored lights.

DAY 1

ACTIVITY

1. Total Darkness

Introduction to the nature of light and darkness.

Being in total darkness is a powerful and memorable experience. If it is at all possible, have students experience the darkness in their classroom because the sensations and emotions they will have in the dark will be most clearly experienced if they are in a comfortable and familiar place. If you cannot black out the classroom—which is a huge challenge—make sure that the place where the students experience darkness allows them to stand or sit comfortably so that they will not be crowded or distracted.

It's important to check the room for any little bit of light before conducting this activity.

MATERIALS

heavy black plastic sheeting
duct tape
candles
matches
1 large sheet of paper

PROCEDURE

Black out the room by covering all windows and doors with heavy plastic. (Doors may need plastic on the outside as well as the inside of the door frame.) At a time when the sun is shining brightly, check to make sure that there are no chinks through which light might pass. Even the smallest bit of light will compromise the experience of being in total darkness.

Create a semicircular screen from a large piece of paper and set it upright on a desk. Set up several candles behind the screen and position the screen so that the students will not see the candles when they are lit.

Instruct the students that when they are in the dark they should ask themselves:

What are my sense impressions?

Which senses are heightened when I am not using my sense of sight?

What are my emotions and thoughts?

What do I experience in the darkness with my eyes open?

What do I experience in the darkness with my eyes shut?

When the class is ready, shut off the lights. The students should sit quietly and experience the darkness. After a couple of minutes, light a match. The students' eyes will be drawn to the light, but they will probably not see much else before the match burns out. After the match has burned out, allow the students to remain in darkness for a little while longer.

Light a candle behind the screen and have students look around the room. Light more candles one by one until there is enough light for the students to be able to see all the objects in the room.

Slowly extinguish the candles one by one. After the last candle is out, allow the students to sit in darkness for a little while longer before turning on the lights.

COMMENT

As the world of the classroom is slowly revealed by light, students may have the experience of a sunrise and experience feelings such as hope or joy. As the candles are extinguished and the light fades, students may have the experience of a sunset and may feel sad or lonely.

This demonstration can also be performed with a small light bulb and a dimmer. The light from the bulb will flicker less than the candles, but a lamp's light is also less "alive" and will not give illuminated objects the same sense of depth.

DAY 2

DEMONSTRATION

2. The Unrevealed Revealer

CHARACTERISTIC: Light can only be seen at its source or by what it illuminates.

Light is the great revealer that opens the world to our gaze. Yet this great revealer never reveals itself, for light itself cannot be seen; we can only see the source of light and what is illuminated by light.

MATERIALS

1 dust-free room

1 projector or strong flashlight

1 tube of heavy black paper

PROCEDURE

Dust the room the night before and mop the floors. Extend the projector's or the flashlight's opening with a tube of black paper to focus the light into a narrower beam. Set some kind of barrier so that the projector or flashlight is hidden from view.

Project the beam of light across the room. If there is little dust, the students will not be able to see the light, only the illuminated area on the wall. Have the students look through the light at a dark background such as a blackboard.

Move a light-colored object into the light's path. If the beam is well focused, you should be able to bring the object quite close to the beam while it remains invisible until it intercepts the beam.

ALTERNATE DEMONSTRATION

2A. The Black Box

CHARACTERISTIC: Light can only be seen at its source or by what it illuminates.

If it is difficult to prepare a blacked-out or dust-free room, the phenomenon illustrated in the previous demonstration can be shown on a smaller scale.

MATERIALS

1 cardboard box painted flat black inside

1 cardboard or heavy-paper tube

black tape

1 flashlight or projector

1 slip of paper

PROCEDURE

Prepare the box by cutting circular holes in both ends of the box. Insert the tube about 3 inches into one of the holes. Seal the joint between the tubing and the box with black tape. The tube opposite the hole is where the light source will be placed. The tube should allow the light projected through the box to go through the hole and not shine on the interior wall of the box. Cut a slit in the top of the box large enough to insert a slip of paper. Cut a rectangular viewing hole in the center of the front of the box.

Shine the light through the box and out of the second circular hole so that the wall is illuminated. Have students look through the viewing hole in the front of the box. It will not be possible to see the beam of light in the box itself. To show that the light is passing through the box, insert a slip of paper into the top slit. The beam no longer shines out of the tube, but the paper is lit.

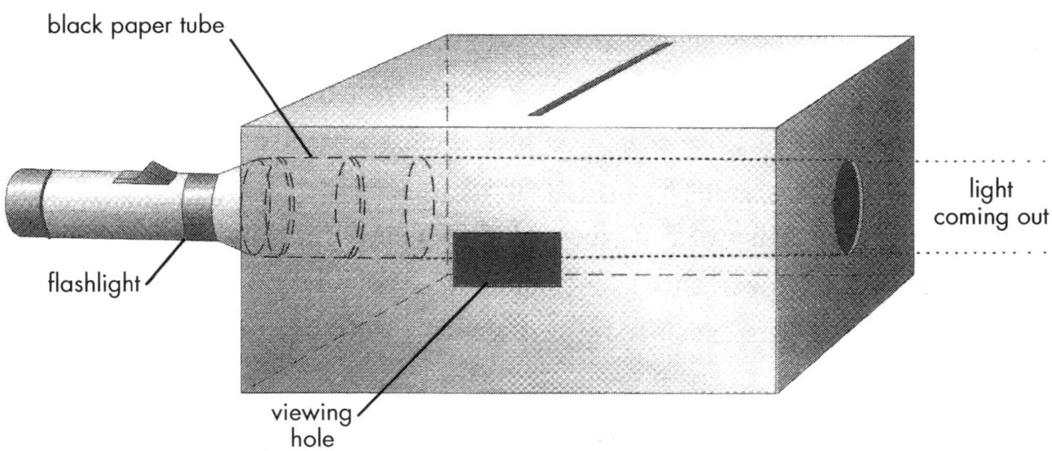

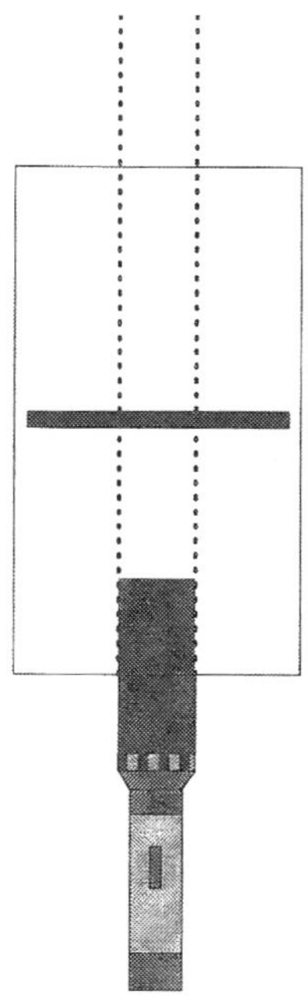

view from top

COMMENT

This demonstration is most effective if the room is not totally blacked out. If there is enough light in the room, students looking through the viewing hole will not be able to tell whether the light is on or off unless a paper or another object is put in the light's path.

DEMONSTRATION

3. The Straight and Narrow Path

CHARACTERISTIC: Unless obstructed, light travels in straight lines in all directions from its source.

MATERIALS

1 strong flashlight, projector, or laser pointer

2 chalk erasers or pads of cloth impregnated with chalk dust

3 mirrors

PROCEDURE

In a darkened room, shine the light across the room and have a student lightly clap the erasers to reveal the path of the light. Have another student stand centered near one wall of the room holding a mirror. Shine the light into the mirror from an angle and have the first student create more dust. The path of the light will be revealed and the students will see that the light is reflected to another point in the room. Have three students stand centered on three walls of the room. Using the mirrors they can reflect the light from one mirror to another and back to the source while the chalk reveals the path of the light. (***CAUTION:*** *If using a laser pointer, make sure that the beam does not shine in anyone's eyes.*)

DEMONSTRATION

4. Transparent, Translucent, Opaque

CHARACTERISTIC: Materials are variously permeable to light: some reflect light; some allow light to pass through; some obstruct light.

PROCEDURE

Black out the room enough to stop most of the light coming in the windows. Cut an opening in one part of the blackout sheet so that light can pass through it. The opening should be a bit smaller than the size of the papers.

Hold the piece of glass in front of the opening. Students should notice no difference in the amount of light coming into the room, and they can easily see what is outside through the glass.

Now hold the tracing paper in front of the opening. Some light still passes through the paper, but it is no longer possible to see out the window.

Hold the piece of heavier paper in front of the opening. It allows only a small amount of light to pass through it.

Do the same with the cardboard; it blocks out the light entirely.

COMMENT

If blacking out the room is too difficult, this demonstration can also be done by holding the three types of paper in front of a bright light source and seeing how much of the light shines through them.

MATERIALS

- heavy black plastic sheeting
- duct tape
- 1 piece of glass
- 1 piece of tracing paper
- 1 piece of heavy paper
- 1 piece of heavy cardboard

DAY 3

DEMONSTRATION

5. The Generation of Color

CHARACTERISTIC: Colors arise from the interplay of light and darkness.

According to Goethe's color theory, colors are created by the interplay of light and darkness. Yellows and reds arise when we look at light through a dark medium, and blue and purple arise when we look at darkness through a light medium. (For anyone not familiar with Goethe's color theory, Heinrich O. Proskauer's The Rediscovery of Color *gives a clear explanation.)*

MATERIALS

a fish tank
black paper
tape
water
milk
1 strong flashlight or spotlight in clamp light fixture

PROCEDURE

Prepare the fish tank by covering the outside back wall with black paper. Set the fish tank where the sun will shine into it at an angle from above. Fill the fish tank about three-quarters full of water. When the students are ready, add a few drops of milk.

If the students stand so that they are looking at the black wall of the fish tank, they will see that the water appears to be blue or purple.

As more milk is added, the color of the background will appear to go from purple to blue. Looking at the background from the side allows the color to be seen more clearly.

Place the light at one end of the fish tank and have students look at the light through the water. It will appear yellow or orange.

COMMENT

This demonstration can also be done on a small scale with a water-filled jar to which a few drops of milk have been added. Whatever the scale, the important variable is the angle at which the light enters the water.

The color red can also be generated with a strong flashlight or spotlight and pieces of white paper. When the papers are placed in front of the light, we see the light shining through the paper change from white to yellow to orange to a rusty red.

The students can now be led to the recognition of where these colors are found in nature. When we see the red of the sunrise or sunset, we are looking at the light of the sun through the layers of the atmosphere. When we see the blue of the sky, we are looking at the darkness of space through the atmosphere lit by the sun.

In the eighth grade, students will learn how the scattering of light makes the sky appear blue, but they need not be introduced to this complex explanation at this time.

DEMONSTRATION

6. Creating Colors with a Water Prism

CHARACTERISTIC: Colors can originate when light passes through a prism.

PROCEDURE

Set up the water-filled fish tank so that the sunlight enters it at an angle, or shine the bright light at the surface of the water at a slant. The light reflecting from the water creates small spectra on the ceiling. When students look through the sides of the tank at the light source and its internal reflection, they will see that the light source is ringed with the colors of the spectrum.

The same demonstration can be done with a plane mirror set in the water, leaning at a slant against the fish tank wall. Move the mirror so that the prismatic colors shine on the ceiling.

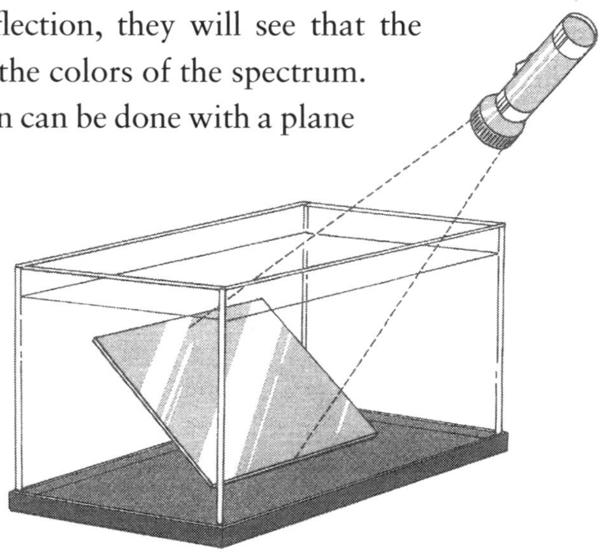

MATERIALS

1 fish tank

water

1 small mirror

bright sunlight or a focused light source

COMMENT

A water prism that works on this principle can be purchased from a science supply company.

ACTIVITY

7. Working with Prisms

CHARACTERISTIC: Colors can originate when light passes through a prism.

This activity could be done during the review on the day following the generation of color.

The glass prism offers students a dramatic experience of seeing colors generated purely through the interaction of the light and the glass. The prism is a mysterious phenomenon that is hard to understand. For sixth graders, it is enough to be able to experience and describe the characteristics of the prism. They can return to this subject in eighth grade and come to understand the phenomena in terms of refraction.

MATERIALS

glass or Lucite prisms

PROCEDURE

Students should look through prisms and take note of where they see colors arising. They may need to be guided to be accurate in their observations.

COMMENT

In their notebooks, students can draw or paste black shapes. Then they can use pencils to color around the edges of the shapes to represent what the shapes look like when seen through a prism.

DAY 4

DEMONSTRATION

8. Mixing Liquid Pigments

CHARACTERISTICS:
Colors may be grouped into primary and secondary colors.
Primary colors can be mixed to form secondary colors.

Although this is a very simple demonstration, it is one of the most beautiful and dramatic ones that can be done during the study of light. In order to achieve the best results, the strength of the pigments should be adjusted beforehand to achieve the best shade of the secondary color, and the demonstration should be done slowly.

PROCEDURE

Mix the red, yellow, and blue watercolors separately in three small jars. Fill the fish tank and the large jars almost full with water. Illuminate one of the large jars by the light from the side or from above. Drop a thin stream of one of the watercolors into the water. When the swirling of the first color has stopped, add a thin stream of another primary color into the colored water.

Repeat this process with the other combinations of primary colors in the three large jars. The three primary colors can also be mixed together in the fish tank by working very slowly and carefully. The final result will be like a very beautiful three-dimensional painting.

MATERIALS

- 1 fish tank
- 3 large jars
- water
- red, yellow, and blue liquid watercolor paints
- 3 small jars or cups
- a light source

COMMENT

There are two sets of primary and secondary colors: the primary and secondary colors of pigments that we find in painting and common usage, and the primary and secondary colors of light. The colors formed by dyes and pigments are called subtractive colors, while colors formed by colored light are called additive colors.

DEMONSTRATION

9. Mixing Light

CHARACTERISTICS:
Colors may be grouped into primary and secondary colors.
Primary colors can be mixed to form secondary colors.

The primary colors of pigments are red, yellow, and blue, but the primary colors of light are red, green, and blue. These can be mixed to create the secondary colors of magenta, yellow, and cyan. Before creating each of the mixtures, students should predict what they expect the mixtures to yield.

In preparation for this demonstration, make sure that the light sources are attached firmly and that the dimmers are labeled. Test the light mixtures carefully before doing this demonstration so that the secondary colors are as pure as possible.

MATERIALS

- a darkened room
- 3 clamp light fixtures
- 3 dimmers for the lights
- red, blue, and green spotlights
- 1 screen or large piece of white paper

PROCEDURE

In a darkened room, shine the red light onto one side of the screen. Slowly bring up the blue light onto the other side of the screen. Where the lights overlap, they will form magenta. Repeat with the blue and the green lights to create cyan. The mixture of the red and green will be the most surprising—yellow. Mixing all three colors together yields a grayish white light.

DAY 5

DEMONSTRATION

10. Colored Afterimages

CHARACTERISTIC: We can experience afterimages and complementary colors.

The study of light might end with a mysterious and profound experience: the physiological colors we create. This experience helps students remain poised and balanced between the subjective and objective realms. One danger of studying physics is that students will reject or disqualify anything that is not physical—anything that cannot be counted, measured, or weighed. Equally dangerous is the possibility that they will develop an attitude that the world has no reality except what we experience within ourselves. The experience of a phenomenon such as colored afterimages helps keep them open to both possibilities.

If you stare at a colored object for a while and then either close your eyes or look at a white surface, you will see a colored afterimage of the object. The afterimage will be the complementary color of the original color. For example, the afterimage of red is cyan, the afterimage of green is magenta, and the afterimage of blue is yellow.

Afterimage demonstrations tend to get the students very excited, so you need to limit how many are done at a time.

PROCEDURE

Mount a colored circle or square on a white background and put a sheet of white paper next to it. Illuminate the colored circle with the bright light. Have the students stare at the colored circle. If they stare long enough, they will begin to see a luminous border of another color surrounding the piece of colored paper. Now have them shift their gaze to the white paper. A bright afterimage of the complementary color of the same size and shape of the colored circle should be visible.

MATERIALS

large paper circles or squares of red, yellow, blue, purple, orange, and green

a bright light

white paper

Repeat the procedure for the various colors.

To do this on a smaller scale, use small squares of the primary and secondary colors. Have the students set a small colored square on a piece of white paper and stare at it. Once the luminous border appears, they can flick the square aside (or cover it with another piece of white paper) while not shifting their gaze. They will see a brightly colored square in the spot where they are looking.

As a variation, students can look at two or three different colored squares overlapping or in an interesting arrangement.

COMMENT

Some students have difficulty seeing the afterimage. Suggesting that they keep their eyes slightly unfocused might help. If a student still cannot see the afterimage, try creating a computer image of a colored square on a white background and have the student stare at it. After 30 seconds, remove the image so that the student continues looking at the empty screen. The light from the screen seems to create more powerful afterimages.

ACTIVITY

11. Afterimages from Afterimages

CHARACTERISTIC: We can experience afterimages and complementary colors.

MATERIALS

brightly colored construction paper cut into 1-inch bands or squares

2 pieces of white paper

PROCEDURE

Make a grid out of bands or squares of colored construction paper on a piece of white paper so that it looks like a checkerboard. Have students stare at the squares until the complementary color becomes visible in the white lines between the squares. After looking at the squares for a minute or two, have them look aside at a white piece of paper. They will see very pronounced colored crisscrossing lines formed by afterimages of the colored squares and afterimages of the bright afterimages.

COMMENT

A variation on the grid can be done by cutting increasingly larger circles out of colored paper and white paper and laying them over each other to create concentric colored rings alternating with white rings. As one stares at the rings, the white circles will be brightly colored by the complementary color.

These experiences reinforce the students' recognition that our experience of color does not only depend on the colors we see but also the colors we create within ourselves.

DAY 6

DEMONSTRATION

12. Colored Shadows

CHARACTERISTICS:
Shadows are created by an obstruction of light.
Colored shadows can be created by a mixture of colored lights.

Just as the Chladni plate was a beautiful way to end the week of studying acoustics, colored shadows are a very fitting end to studying light. I recommend that this topic be presented after the review of afterimages on the first day of the next week.

PROCEDURE

Prior to the demonstration, set up the light fixtures at a distance of six to eight feet away from the screen. Arrange them to be three to four feet apart from one another so that they will cast shadows that are well separated. Make sure that the light sources are attached firmly to chairs or desks. Set the object between the lights and the screen so that its shadow appears on the screen.

MATERIALS

- a darkened room
- 4 clamp light fixtures
- dimmers for the lights
- red, blue, green, and white spotlights
- 1 object such as a vase
- 1 screen or large piece of white paper

To demonstrate colored shadows, begin with the green light. Students will notice a green background, and the object's shadow will be black. Diminish the green light and increase the blue light. Students will now see a blue background and a black shadow where the blue light is blocked. Slowly increase both the blue and green lights. The background becomes cyan and now there are two shadows: a green one where the blue light is blocked and a blue one where the green light is blocked. Repeat with blue and red lights and with red and green lights.

Now tell the students that they are going to experience a mystery. Turn up the red light and students will see the red background and a black shadow as they did before. Slowly increase the white light. A red shadow appears where the white light is blocked, and a cyan shadow appears where the red light is blocked. Repeat with blue and white; these create blue and yellow shadows. Repeat with green and white; these create green and magenta shadows.

Now put on all three colored lights, turn up the white light, and create a spectrum of colored shadows.

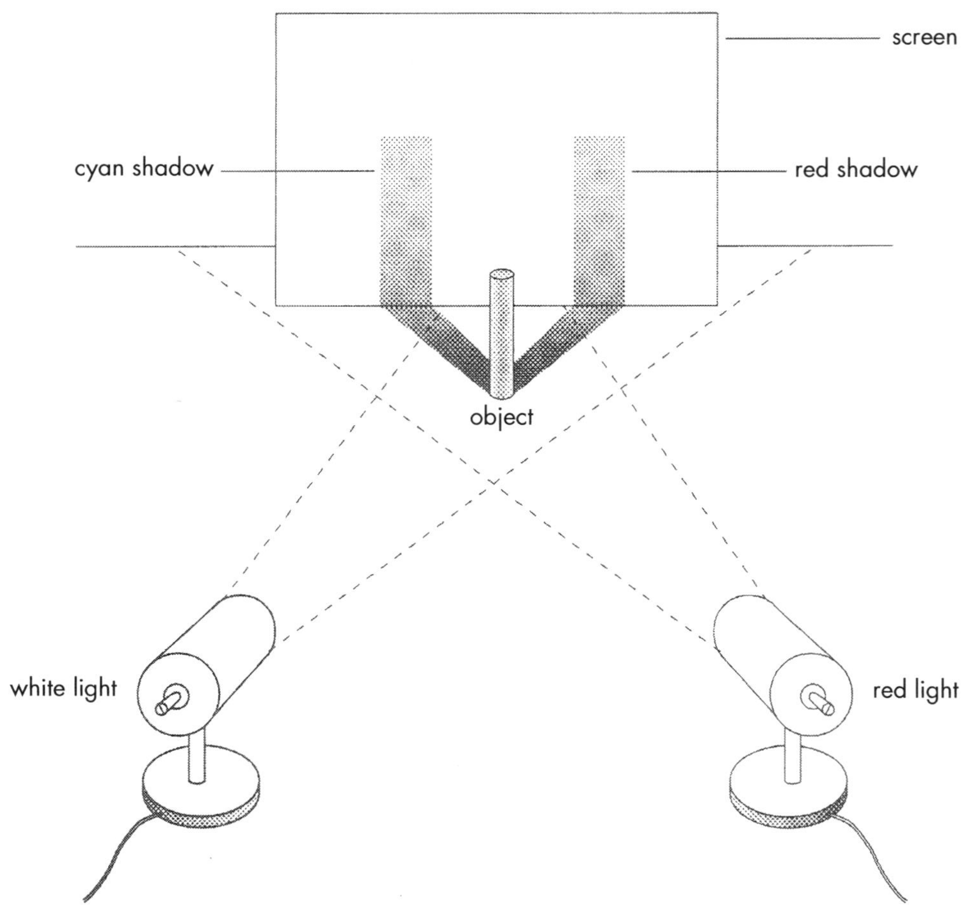

COMMENT

Where did the colors in the second set of demonstrations come from? How is it possible for a red light to create a cyan shadow, a blue light to create a yellow shadow?

If we were able to look at a colored shadow alone and block out the surrounding colored light, the shadow would look grey. But if we look at the shadow in the context of the surrounding light, it looks colored. So is the shadow colored or not? If it looks colored to us, does that mean that it *is* colored? Such questions can lead to discussions that highlight the importance of the observer's experience in the process of scientific inquiry.

ADDITIONAL & ALTERNATIVE ACTIVITIES

ACTIVITY

13. Orientation in the Darkness

I do not suggest doing this on the same day as the students experience total darkness, but it is a challenging and fun activity to do while the room is still blacked out.

PROCEDURE

Place a bell somewhere in the room, point out its location to the students, and have a student stand by the door. Shut off the lights and let the student try to make her way from the door to the place where the bell is sitting. When the student has found her way to the bell and has rung it, turn on the lights and let someone else try to navigate in the dark.

MATERIALS

a blacked-out room
1 bell

COMMENT

Through this activity students may realize that they can "see" in the dark with the help of their sense of touch and the "light" of their memory.

ACTIVITY

14. Light and Distance

CHARACTERISTIC: The intensity of light decreases with distance.

MATERIALS

1 candle
matches
8 pieces of stiff 8½"x11" white paper
1 yardstick
1 thumbtack
1 roll of masking tape

PROCEDURE

Set a candle on a desk in the center of the room. Cut one of the pieces of paper into an 8-inch square and fasten it to the end of the yardstick with a thumbtack. Fold the rest of the papers so that they make 8-inch squares with 3.5-inch flaps. Tape the 3.5-inch flaps to desks so that the papers will stand upright, and arrange the desks so that the papers form a ring 3 feet from the candle. Have students stand in a circle around the desks, darken the room, and light the candle.

Students will be able to observe that the papers are all lit to the same degree. Hold the yardstick with the square of paper on its end in different places 3 feet above the candle. Except for the shadow of the yardstick, that piece of paper will be illuminated with the same intensity as the ring of papers.

ACTIVITY

15. The Intensity of Light

CHARACTERISTIC: The intensity of light increases with distance.

PROCEDURE

Set the candle in the middle of the room. Have four students stand in a ring around the candle a given distance away. Four more students should stand at twice the distance, and four more at three times the distance. Each student should hold a piece of paper and face the candle. When the room is darkened and the candle lit, students will see that the illuminated papers show how the intensity of light decreases.

MATERIALS

12 pieces of stiff white paper, each about 12 inches square

1 candle

matches

DEMONSTRATION

16. The Laws of the Shadow

CHARACTERISTICS:

Shadows are caused by an obstruction of light.

The darkness of the shadow is determined by the intensity of the light.

The shape of a shadow bears a direct relationship to the shape of the object casting the shadow.

A good deal of time can be spent in the study of shadows, but this need not necessarily be done during the physics block. In sixth-grade art classes, black and white drawing is introduced, and some of the topics will include working with shadows. Some of the following demonstrations and activities can be done during these classes. Others can be assigned as extra projects.

MATERIALS

- 1 strong light source
- 1 weak light source
- 1 light-colored table or 1 table covered with white cloth or paper
- a variety of objects of different shapes

PROCEDURE

Set one of the objects on the table and shine the strong light on it from one side at an angle. Shut that light off and shine the weak light from the same distance and at the same angle. Students will be able to see how the darkness of the shadow is determined by the intensity of the light but that the size of the shadow does not change.

Arrange the series of objects on the table and shine the strong light on them at a 45° angle. Students will observe that the shape of a shadow bears a direct relationship to the shape of the object casting the shadow.

Hold the light at the same angle and move it closer to the table. Observe how the shadows change. Now move the light more directly overhead and look at the changes. Students will easily see that a shadow's shape changes according to the position of the light.

Move the light closer to and farther from one of the objects. The students will see that a shadow can be focused by moving the light source closer to or farther from the object.

Hold two lights sources so that they shine on an object from different angles. This shows how multiple shadows are created.

ACTIVITY

17. Observing Shadows Cast by Sunlight

PROCEDURE

Take a walk on a sunny day when there are plenty of shadows being cast. Observe the relationship between an object and its shadow. When walking among trees, see if students notice that, regardless of the shape of the openings between the leaves of a tree, the shape of the patch of light on the ground is always round—the shape of the sun.

ACTIVITY

18. Shadow Play

PROCEDURE

Hang a white sheet and strongly illuminate it from behind. Students can make hand shadows and shadow puppets or perform a shadow play.

MATERIALS

1 white sheet

1 strong light source

ACTIVITY

19. How Colors Look in Colored Light

CHARACTERISTIC: Our perception of the color of an object depends on the color of the light illuminating it.

PROCEDURE

Darken the room and project a red light. Put different colored papers or objects in the path of the light to see what colors they appear. Repeat with other colored lights.

MATERIALS

clamp light fixtures

colored spotlights

colored papers or objects

DEMONSTRATION

20. The Bentham Disk

CHARACTERISTIC: Colors arise from the interplay of light and darkness.

MATERIALS

1 Bentham disk (see diagram)
poster board
1 push pin

PROCEDURE

Make a Bentham disk by photocopying the accompanying diagram and pasting it onto poster board. Insert the push pin in the center, hold the pin with the point up, and spin the disk. As the disk spins, faint colors are visible on the edges of the black and the white areas. Different colors appear depending on the direction of the spin.

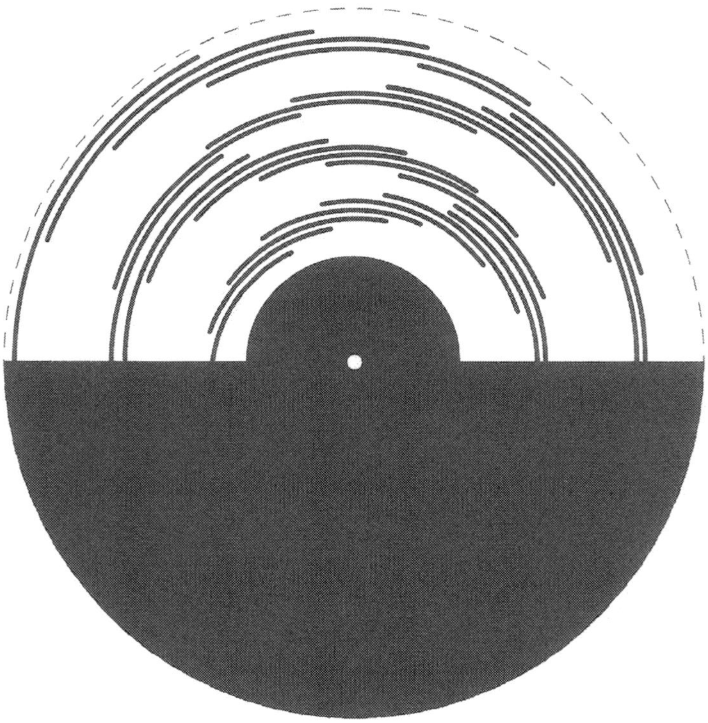

ACTIVITY

21. Making Simple Chromotropes

CHARACTERISTIC: Primary colors can be mixed to form secondary colors.

PROCEDURE

Mark several 3-inch circles on the poster board and cut them out. Shade the circles in two or three colors on each side. (See diagram.) The colors can be combined in various types of designs. Use the compass point to make two holes ¾ inch apart in the center of the circle. Thread a 3-foot piece of string through the holes and tie the ends of the string together. Position the chromotrope at the midpoint of the resulting loop.

To spin the chromotrope, hold each end of the string loop and gently twirl the disk away from you. When the string is wound fairly tight, pull both hands apart and the disk will begin to twirl. Alternately pull and slightly release the string as the disk nears the end of the twirl and it will continue to spin.

MATERIALS

pencil compass
white poster board
scissors
crayons or magic markers
string

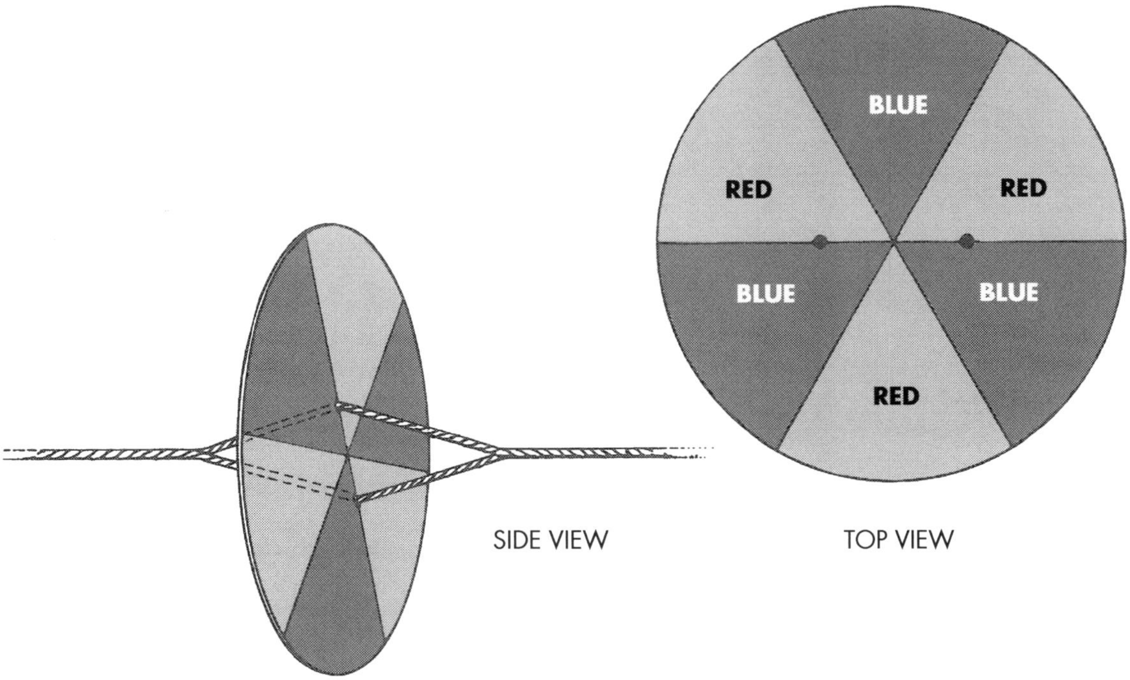

SIDE VIEW TOP VIEW

Notes

The Study of Heat in the Sixth Grade

In all of our physics studies, we begin with the human being. Heat and cold affect us deeply, and the expansion and contraction that we experience in our bodies and souls when we experience hot or cold are similar to the behavior of solids, liquids, and gases when they are heated and cooled. For this reason, it is meaningful to students to see their inner experiences mirrored and objectified in the outer world.

With the study of warmth and cold, our work in physics relies more on external than internal experiences. The sense organs for sound are buried deep inside our skulls and the sense organs for light are close to the surface, but the sense organs for warmth and cold are spread out all over the surface of our bodies.

We want students to keep the broadest possible view of physical forces and not rely only on their physical experiences. For this reason we will begin our study of heat by considering the less tangible aspects of warmth: the warmth of colors, of sounds, and of materials. While this might seem overly subjective and possibly unscientific, it is essential to a broad and encompassing approach to learning about the physical world.

Sounds affect substances relatively little, light slightly more so, but heating and cooling can change a substance's physical state and chemical composition. Some of these changes will be examined in this block; others will wait until the study of chemistry in seventh and eighth grades.

We will explore the sources of heat. Then, having explored how heat originates through combustion, light, friction, stress, chemical reactions, and electricity, we will examine the effects of heat. The final topic for the study of heat in the sixth grade is how heat travels through various media.

If we think metaphorically, we might see that the three ways heat travels—conduction, convection, and radiation—can be related to the human faculties of willing, thinking, and feeling.

Conduction, the most physical of the three processes, requires a solid or a liquid to transfer heat. In conduction, each part of the substance heats the adjacent part, passing the heat to the next part. We can see this in relation to our own will and how our actions have direct consequences.

Radiation, the least physical of these processes, requires no physical contact or medium whatsoever; radiant heat travels through gases—the most refined media—and can even travel through a vacuum. This can be seen in relation to our thinking, which is the least physical of our capacities.

Convection lies between these two poles. With its moving and weaving quality, with its polarities of hot above and cool below, heat travels through convection in liquids and gases. Convection can be compared to our capacity for feeling, which bridges the physical and the spiritual.

Such thoughts are not meant to be shared with the children, but they suggest ways that we, as teachers, can look at a physical force from a broader perspective so that we may continue to seek new meaning and unveil new mysteries.

TOPICS

In the sixth grade, the study of heat deals with three major topics:

- The sources of heat
- The effects of heat
- How heat travels

CHARACTERISTICS

In working with these topics, the following characteristics will be revealed:

- Heat originates from combustion, electricity, chemical (and nuclear) reactions, friction, and light.
- Heating and cooling affect a substance's physical state.
- Heating causes expansion, and cooling causes contraction in most solids, liquids, and gases.
- Heat travels through solids and liquids by conduction; through liquids and gases by convection; and through gases by radiation.

These topics and concepts can be revealed by the following demonstrations and activities:

DAY	DEMONSTRATIONS AND ACTIVITIES	CHARACTERISTIC
1	■ The Sources of Heat	■ Introduction to the nature of warmth and cold. ■ Heat originates from combustion, electricity, chemical (and nuclear) reactions, friction, and light.
2	■ The Effects of Warmth and Cold ■ The Ball and Ring ■ The Stretching Wire	■ Heating and cooling affect a substance's physical state. ■ Heating causes expansion and cooling causes contraction in most solids.
3	■ The Expansion of Liquids ■ The Expansion of Gases	■ Heating causes expansion, and cooling causes contraction in most liquids and gases.
4	■ Conduction in a Metal Rod ■ A Poor Conductor ■ A Good Conductor ■ A Candle's Heat ■ Through a Glass Brightly	■ Heat travels through solids and liquids by conduction and through gases by radiation.
5	■ Visible Air Currents ■ Convection Currents ■ Hooray for the Red, White, and Blue	■ Heat travels through liquids and gases by convection.

Additional or alternative activities are included at the end of this chapter.

DAY 1

INTRODUCTION: WARMTH AND COLD

Do we experience warmth and cold only through our sense of touch? We speak of the warmth of colors, the warmth of sounds, the warmth of materials, and the warmth of our voices. What do we mean? In preparation for studying heat, ask students to consider whether colors, sounds, or materials seem warm or cool. Although these observations will be somewhat subjective, they help the students think of warmth and cold as more than simply physical properties.

DEMONSTRATION

1. The Sources of Heat: Combustion

CHARACTERISTIC: Heat originates from combustion

Burning is the most common source of heat. The many changes wrought by combustion will be studied extensively in the seventh- and eighth-grade chemistry main lesson blocks. In the following demonstrations, students should simply experience the process of burning as a source of heat.

MATERIALS

- 1 aluminum tray or pie plate
- sand
- 1 watch glass or small flameproof dish
- matches
- paper
- isopropyl or other type of alcohol

PROCEDURE

Prepare a sand tray by putting about 1 inch of sand in an aluminum tray or pie plate. Burn a piece of paper in the sand tray and have students observe how burning releases both light and heat. Light a little bit of alcohol in the watch glass and have students observe and feel the warmth of the flame.

COMMENT

When reviewing this demonstration, compare different combustible sources of heat in terms of heat emission, thoroughness of combustion, and byproducts.

DEMONSTRATION

2. The Sources of Heat: Electricity

CHARACTERISTIC: Heat originates from electricity.

PROCEDURE

Place the steel wool atop the sand in the aluminum tray. Touch both terminals of the battery to the steel wool strands. The steel wool will begin to glow and will combust briefly.

MATERIALS

1 aluminum tray or pie plate
sand
1 9-volt battery
small strands of steel wool

DEMONSTRATION

3. The Sources of Heat: Chemical Reactions

CHARACTERISTIC: Heat originates from chemical reactions.

Many chemical reactions are exothermic—i.e., they release heat. When working with chemicals, wear appropriate eye and hand protection.

PROCEDURE

Pour about a teaspoon of lye into a beaker half full of water and stir. Hold the beaker in place while students touch the base to feel the heat created by the reaction. (***CAUTION:*** *Put the lye into the water—do not pour the water into the lye.*)

COMMENT

You can also demonstrate how a mixture can create cold by chemical means by mixing a tablespoon of ammonium nitrate into a cup of water. These chemicals can be disposed of safely by diluting and flushing down the sink.

MATERIALS

a small amount of lye (sodium hydroxide)
1 100-ml or 250-ml beaker
1 chemical scoop or teaspoon
1 stirring rod
protective goggles
protective gloves
water

DEMONSTRATION

4. The Sources of Heat: Friction

CHARACTERISTIC: Heat originates from friction.

Students may have encountered this phenomenon in woodworking class, but it is still worth drawing their attention to it more consciously.

MATERIALS

1 saw or drill with a ¼-inch drill bit

1 piece of wood

PROCEDURE

Begin sawing or drilling into a piece of wood and after a little while have students feel the saw blade or drill bit.

ACTIVITY

5. The Sources of Heat: Stress

CHARACTERISTIC: Heat originates from friction.

When materials are subjected to forced movement, internal stresses can cause them to become hot enough to fracture. In this simple demonstration, students will see how a piece of steel can be broken by repeated bending. This process generates a fair amount of heat.

MATERIALS

Pieces of metal coat hanger, 8–12 inches long

PROCEDURE

Sand or file any sharp ends of the hanger or cover them with masking tape. Have each student hold the ends of the pieces of the coat hanger and force them to bend in the middle. Continue bending the hanger repeatedly until it breaks. Touch the spot where the hanger broke; it will be very hot.

ACTIVITY

6. The Sources of Heat: Light

CHARACTERISTIC: Heat originates from light.

This common phenomenon still has the power to delight students, even if they have done it before. It may be best done outdoors during recess when students can try to ignite various materials.

PROCEDURE

Put the piece of paper or leaves in the sand-filled tray and focus the sun's rays on them with a magnifying glass until they begin to smolder. Students can also try focusing the sun's rays on their palms to see how the spot quickly gets too hot to endure.

MATERIALS

magnifying glasses

thin pieces of paper or leaves

1 aluminum tray or pie plate

sand

DAY 2

DEMONSTRATION

7. Warming and Cooling

CHARACTERISTIC: Heating and cooling effect a substance's physical state.

Warmth is a great "unshaper," releasing the form of almost anything that it touches. Cold is a "form-giver," restoring shape to that which was fluid.

MATERIALS

- 1 block of wax or paraffin
- 1 large spoon or small pan
- 1 alcohol lamp or hot plate
- 1 metal tray
- 1 small container

PROCEDURE

Melt some wax or paraffin in the spoon or pan over an alcohol lamp or the hot plate. Pour some of the melted wax onto a metal tray. Pour some in a small container and allow some to remain in the pan. Students will be able to observe how the wax loses its shape as it is heated and regains it as it cools.

COMMENT

The same demonstration can also be done with chocolate, though you may need to melt the chocolate in a double boiler so that it doesn't scorch.

During the review, students might mention that heating some substances—like eggs—causes them to solidify. Students will explore the nature of proteins that react to heat by solidifying during the study of foodstuffs in eighth-grade chemistry block.

DEMONSTRATION

8. The Ball and Ring

CHARACTERISTIC: Heating causes expansion and cooling causes contraction in most solids.

This is the classic demonstration of the expansion of solids by heating. In the review it would be good to give examples of how this principle is utilized in industry and everyday life. A ball and ring apparatus needs to be bought from a science supply company, but it is well worth the expense because it illustrates this phenomenon so clearly.

MATERIALS

- 1 ball-and-ring apparatus
- 1 Bunsen burner or propane torch
- matches or lighter

PROCEDURE

If the ball fits through the ring when cool, heat the ball and demonstrate that it no longer fits through the ring. If the ball doesn't fit through the ring when cool, heat the ring and demonstrate how the ball fits through easily.

DEMONSTRATION

9. The Stretching Wire

CHARACTERISTIC: Heating causes expansion and cooling causes contraction in most solids.

This demonstration can be done in addition or as an alternative to the ball and ring.

Metal cables, girders, and railroad tracks all expand when they are heated. This property is taken into account in engineering and construction.

PROCEDURE

Attach the weight to the wire and attach the wire to the stand. Raise the ring support so that the weight hangs about a quarter-inch above the table top. Heat the wire with the torch or burner, moving the flame up and down the length of the wire. As the wire is heated, it will stretch, and the weight will eventually touch the table. Allow the wire to cool and observe how the weight rises back to its original position.

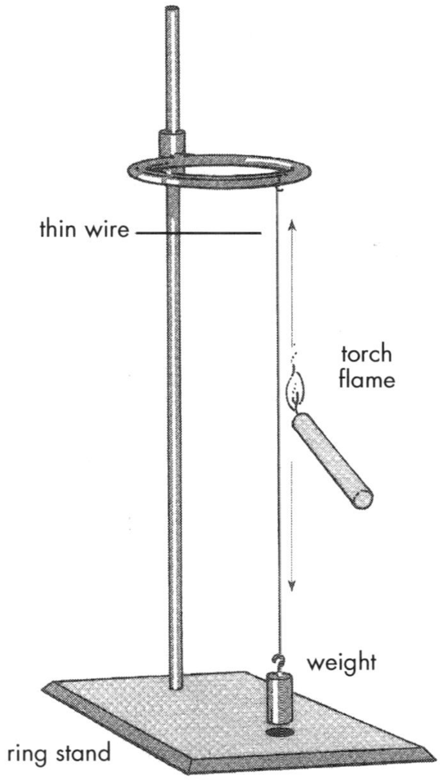

MATERIALS

1 weight, either 250 grams or 500 grams

1 laboratory stand with a ring clamp

1 piece of thin uninsulated copper wire

1 propane torch or Bunsen burner

matches or a striker

COMMENT

A bimetallic strip, which can be purchased from a science supply company, allows you to demonstrate the different expansion rates of different metals.

DAY 3

DEMONSTRATION

10. The Expansion of Liquids

CHARACTERISTIC: Heating causes expansion and cooling causes contraction in most liquids.

MATERIALS

- 1 laboratory stand
- 1 ring clamp and wire gauze
- 1 Erlenmeyer or round-bottomed flask
- 1 one-holed rubber stopper
- 1 piece of glass tubing about 12″ long
- cold water
- food coloring
- an alcohol lamp or Bunsen burner on low
- matches

PROCEDURE

Carefully insert the glass tube through the stopper so that it protrudes about ½ inch below the stopper and 10 inches above it. Fill the flask with cold tinted water and seal the flask with the rubber stopper. As the flask is heated, water will rise in the tube; as it cools, the water will sink in the tube.

COMMENT

Try to ignore the students' pleas to keep heating the water until it shoots out of the glass tube.

After this demonstration, the principles of the thermometer may be examined and discussed. Thermometers can be made in the seventh-grade study of heat.

DEMONSTRATION

11. The Expansion of Gases

CHARACTERISTIC: Heating causes expansion, and cooling causes contraction in most gases.

Like solids and liquids, gases expand when they are heated. Because most gases are invisible, we can only see that they expand by heating them in something that reveals the effects of the heated gas.

PROCEDURE

Secure the flask to the laboratory stand and place it about an inch above the heat source. Stretch the balloon over the top of the flask and then gently heat the flask. As the air inside the flask expands, the balloon will be inflated.

COMMENT

Be prepared for students to giggle as the balloon slowly inflates and eventually stands upright and swells.

MATERIALS

- 1 Erlenmeyer or round-bottomed flask
- 1 ring stand
- 1 ring clamp and wire gauze
- 1 balloon
- 1 alcohol lamp
- matches

DAY 4

ACTIVITY

12. Conduction in a Metal Rod

CHARACTERISTIC: Heat travels through solids and liquids by conduction; through liquids and gases by convection; and through gases by radiation.

Students in the sixth grade have examined how sound travels and how light travels. Now they are ready to learn how heat travels. While sound travels differently through various media, the distinctions are not as obvious as those for heat. With heat, there are three distinct ways in which it travels, depending on the medium.

Most of us have unconsciously learned about conduction. For instance, we know not to stir hot liquid with a metal spoon because it conducts the heat of the liquid too well. Through the following demonstrations, students should become even more aware of the process of conduction—how heat travels through solids and liquids.

MATERIALS

1 copper rod or pipe about 3 feet long

1 Bunsen burner or propane torch

matches or a striker

PROCEDURE

Have a number of students hold the metal rod horizontally between their fingers. The first student should have his or her fingers about 6 inches away from the end that is going to be heated, the next one a few inches further down, and so on. Heat one end of the rod. As the heat travels down the rod's length, one student after another will have to let go because the rod becomes too hot to hold. Students can note the time it takes for heat to travel from the heated end of the rod toward the other.

COMMENT

This demonstration can also be done with small wax balls placed along the length of the rod. As the heat travels along the rod, the balls will fall off.

DEMONSTRATION

13. A Poor Conductor

CHARACTERISTIC: Heat travels through solids and liquids by conduction; through liquids and gases by convection; and through gases by radiation.

Heat travels through liquids both by conduction and convection. Compared to most solids, liquids are poor conductors of heat.

MATERIALS

1 laboratory stand and clamp

1 large test tube

water

an alcohol lamp or low-burning Bunsen burner

matches

PROCEDURE

Attach the test tube to the stand, positioned at a slant so that the top of the test tube can be heated. Fill the test tube three-fourths full of water. Direct the heat at the top of the test tube until the water boils. The bottom of the test tube will remain cool to the touch.

ACTIVITY

14. A Good Conductor

CHARACTERISTIC: Heat travels through solids and liquids by conduction; through liquids and gases by convection; and through gases by radiation.

If a material is a very good conductor, it allows heat to move through it. In this activity, students will see that a good conductor stops the paper from burning by conducting heat away, while the poor conductor allows the paper to heat up to the kindling point.

PROCEDURE

Wrap a strip of paper about an inch wide in a spiral around the middle of each of the rods and tape in place. Hold the part of the rod wrapped in paper over the flame. Because the copper rod conducts heat so well, the paper doesn't burn. Because wood is a poor conductor, the paper wrapped around the wooden rod begins to char. (***CAUTION:*** *Be careful not to let the paper or wooden rod begin to burn.*)

MATERIALS

1 wooden rod and 1 copper rod of similar size

2 strips of paper

masking tape

an alcohol lamp or low burning Bunsen burner

matches

ACTIVITY

15. A Candle's Heat

CHARACTERISTIC: Heat travels through solids and liquids by conduction; through liquids and gases by convection; and through gases by radiation.

We have all observed a candle and felt the heat above the flame. Most of this heat is traveling by convection, i.e., the air above the flame is heated and rises. Through this activity students recognize that the heat of the candle also radiates in all directions, even downward.

MATERIALS
candles
matches

PROCEDURE

Give each student or group of students a candle. Have them light the candles and hold their hands above, below, and to the sides of the flame. While the area above the flame is hottest because of convection, the heat of the candle will also be perceptible on the sides and even below the flame.

DEMONSTRATION

16. Through a Glass Brightly

CHARACTERISTIC: Heat travels through solids and liquids by conduction; through liquids and gases by convection; and through gases by radiation.

Because we are all familiar with the sensation of touching something hot, conduction is a common and direct experience. Unlike conduction or convection, radiant heat is not dependent on a medium to carry it from one place to the other, and it can travel through a transparent substance like glass. This theme will be further examined in the eighth grade during the study of the vacuum.

MATERIALS
1 high intensity bulb
1 clamp light fixture
1 pane of glass

PROCEDURE

Turn on the high intensity bulb and have students stand in a ring around it, holding their hands up to it on all sides to feel the radiant heat. Have the students close their eyes while you pass the pane of glass between their hands and the bulb. Ask students if they can feel any difference in heat when the glass is between their hands and the bulb.

DAY 5

DEMONSTRATION

17. Visible Air Currents

CHARACTERISTIC: Heat travels through solids and liquids by conduction; through liquids and gases by convection; and through gases by radiation.

In convection, the heat is carried from one place to another by a fluid medium such as a liquid or gas. This occurs because the liquid or gas that is heated becomes less dense and rises while the cooler, denser liquid or gas falls. Convection currents are not easy to see, so we must find ways to make them visible.

PROCEDURE

Set the candle in front of the screen so that when the light is on, the candle is in focus. Darken the room, turn on the light source, and light the candle. The candle flame will be seen as a grey shadow and the air currents above it will be visible. When the candle is snuffed, the smoke continues to rise in beautiful convection currents.

MATERIALS

- 1 projector or bright, focused light source
- 1 candle
- 1 screen or large piece of white paper
- matches

DEMONSTRATION

18. Convection Currents in Water

CHARACTERISTIC: Heat travels through solids and liquids by conduction; through liquids and gases by convection; and through gases by radiation.

Rheoscopic fluid has tiny iridescent particles in it that reveal subtle movements in the fluid. It can be purchased from a science supply company and is well worth the expense because it reveals convection so clearly.

MATERIALS

1 large Erlenmeyer or round-bottom flask

1 laboratory stand and clamp

1 ring clamp and wire gauze

rheoscopic fluid

1 alcohol lamp or low-burning Bunsen burner

matches

PROCEDURE

Fill the flask with the rheoscopic fluid and hold the flask in place on the ring clamp. As you gently heat the flask, the students will begin to see the convection currents of the fluid.

DEMONSTRATION

19. Hooray for the Red, White, and Blue!

CHARACTERISTIC: Heat travels through solids and liquids by conduction; through liquids and gases by convection; and through gases by radiation.

This is a fitting final demonstration for the week's study of heat.

MATERIALS

1 fish tank

2 beakers, either 500- or 1000-ml

2 small jars with screw-on lids

red and blue food coloring

1 hot plate or electric kettle

1 small pot

1 medium-sized bowl

ice cubes

water

PROCEDURE

Fill the fish tank with room-temperature water. Fill the beakers with water and place them upside down in the fish tank so that they are about 6 or 8 inches apart.

Mix the ice and water in the bowl to chill the water thoroughly. Fill one of the small jars with cold water, color it intensely with blue food coloring, and screw on the lid. Heat a small amount of water and fill one of the small jars with it. Color the hot water intensely with red food coloring and screw on the lid.

Put the jar with cold water sideways on the upside down beaker so that it faces the front of the fish tank. Hold the jar with one hand

and gently unscrew the lid with the other. Try to gently remove the lid without disturbing the surrounding water too much. When the lid is removed, blue-colored water will begin to sink to the bottom of the fish tank. Repeat with the jar of hot water. When the lid is removed, the red colored water will begin to rise to the top.

COMMENT

This demonstration requires plenty of practice so that the little jars don't fall off the upside down beakers and so that your hands do not interfere with the convection currents. When done properly, a beautiful blue waterfall is created, matched by a rising red plume. Once the small jars are empty, the red, white (clear) and blue bands of water remain visible for quite a while.

ADDITIONAL OR ALTERNATIVE ACTIVITIES

ACTIVITY

20. Warm to the Touch

CHARACTERISTIC: Heat travels through solids and liquids by conduction; through liquids and gases by convection; and through gases by radiation.

Different materials in a room feel warmer or cooler to the touch. This is due to their relative conductivity. Those materials that easily conduct heat draw the heat out of our hands and feel cold; those that are not good conductors feel warmer.

PROCEDURE

Have the students press their palms against the surfaces of various materials in the classroom. They should ascertain which materials feel warmer and which materials feel cooler. If there is no exposed metal in the classroom, bring in a plate of steel or copper. It will feel very cold to the touch.

COMMENT

It is not easy to explain to students why some materials feel warmer and others cooler, even though all the materials in the classroom are likely to be the same temperature. In the seventh or eighth grade, when the conductivity of various substances is examined, students can recall this activity and see if they can explain the relative warmth of materials in terms of their conductivity.

ACTIVITY

21. Cold Hands

CHARACTERISTIC: Heating causes expansion, and cooling causes contraction in most solids, liquids, and gases.

Cold has a hardening, deadening influence on the human body. This activity, or others like it, can impress this fact on the students.

MATERIALS

large shallow pan or bowl
water
ice cubes
yarn

PROCEDURE

Have students place their hands in the ice-cold water for a few minutes, and when they take their hands out and dry them, have them try to perform a task that requires manual dexterity, such as tying their shoes or braiding yarn. Because the cold has made their hands so stiff, the students will have difficulty performing a task that would be simple with warm hands.

ACTIVITY

22. Cracking Marbles

CHARACTERISTIC: Heating causes expansion, and cooling causes contraction in most solids, liquids, and gases.

Sudden cooling can cause materials to break or become brittle. Because glass is such a poor conductor of heat, suddenly cooling the outside or inside of most types of glass containers will cause them to break. Similarly, a heated marble can be fractured by sudden cooling.

PROCEDURE

Heat a marble by holding it in the flame of the burner with a pair of pliers or Vise-Grip. When the marble is glowing red, drop it into a glass of cold water. The sudden cooling will cause the marble to fracture. Students can also do this activity at home on the stove.

MATERIALS

glass marbles
pliers or Vise-Grip
heat-resistant gloves
a strong heat source, such as a stove or Bunsen burner
matches or striker
a glass
cold water

DEMONSTRATION

23. Boiling Water in a Paper Cup

CHARACTERISTIC: Heat travels through solids and liquids by conduction; through liquids and gases by convection; and through gases by radiation.

This demonstration is more on the order of a parlor trick than a scientific experience, but it is a memorable way to end the day's work with conduction. You might choose to reserve it for the seventh-grade chemistry block when students learn about the kindling temperatures of various substances.

MATERIALS

1 unwaxed paper cup that has no bottom lip
3 paper clips
1 laboratory stand
1 ring clamp
water
an alcohol lamp or a low-burning Bunsen burner
matches

PROCEDURE

Hook the paper clips through the top edge of the cup so that the cup can be suspended over the heat source from the ring clamp. Pour about ¾ inch of water into the cup and light the heat source under the cup. Although the paper might singe, it will not burn, and the water will boil.

COMMENT

If students wish to perform this demonstration at home, the cup can be held by kitchen tongs over a low flame on the stove, with a parent on hand to supervise.

ACTIVITY

24. Hot Air Rises

CHARACTERISTIC: Heat travels through solids and liquids by conduction; through liquids and gases by convection; and through gases by radiation.

In popular parlance, we say that heat rises. After working with convection currents in gases, students should become aware enough of the phenomena to be able to make the subtle distinction that it is the hot air *that rises.*

MATERIALS

1 piece of paper
1 pair of scissors
1 needle
thread
1 candle
matches

PROCEDURE

Make a pinwheel or cut a paper spiral from a circle of paper. If using a spiral, pierce a small hole in one end of the spiral and hang it above a lit candle; the spiral will begin to turn. If using a pinwheel, hold it over the flame.

COMMENT

Brass party chimes or holiday carousels may also be used to illustrate this principle. A common misconception is that the heat is rising. Students should be led to understand that the heat has warmed the air, and that the air is rising.

DEMONSTRATION

25. Convection Tube

CHARACTERISTIC: Heat travels through solids and liquids by conduction; through liquids and gases by convection; and through gases by radiation.

A convection tube is a long hollow glass tube that is bent into a closed square shape with an opening on the top. When the water in the bottom of the tube is heated, it rises, forcing the colder water to cycle around the tube. Convection tubes are available from science supply companies.

PROCEDURE

Fill the convection tube with water and add a few drops of food coloring to the opening or sprinkle in a few grains of potassium permanganate. Heat the water by holding the heat source near one corner at the bottom of the tube. As the hot water rises, it displaces the colored water and reveals the convection pattern.

MATERIALS

1 convection tube
food coloring
1 alcohol lamp
matches

Notes

The Study of Magnetism in the Sixth Grade

With the study of magnetism we truly enter the earthly realm, for magnetism occurs in minerals and metals. Our perception of magnetism is largely through our sense of movement and our sense of sight. We can feel the pull of a magnet as it is attracted to something or see the magnet move as it turns to face the north.

When we studied sound, light, and heat, we tried to let the students experience these forces in more than just physical terms. With magnetism—and even more so with electricity—we must strive for the opposite effect. These forces are so elusive and mysterious that it is easy for them to appear magical. If we want to help the students perceive and experience these forces objectively, we must keep our activities straightforward, simple, and practical.

The generation or induction of magnetism is much more difficult for students to understand than the generation of sound, light, or heat. Students can readily perceive how sound, light, and heat are generated, but with magnetism, it is different: a magnet simply *is* magnetic. We cannot and should not expect students to understand how magnetism is created; it is sufficient for them to know how it can be induced and propagated.

We can begin the study of magnetism by telling a story of its discovery. Such a story helps establish and ground the students' first experience of this force in the scope of human history. Then we can introduce the salient properties of magnets: their ability to attract certain metals, their tendency to orient themselves to the north and south, and their powers of attraction and repulsion.

All of these properties are simple to grasp but very difficult to explain. Luckily, most students are content to experience and experiment with magnetism without seeking underlying explanations.

Magnetism can travel through many substances. When students study how magnetism travels, they will note the difference between magnetic force and the forces of sound, light, and heat. Those forces can be slowed, obstructed, or reflected, while magnetism travels through almost any medium without interference.

We complete the study of magnetism by unveiling the mystery of the field of force. When we do this, we are making the invisible visible, and we need to take care not to trivialize the mystery. If we cultivate a serious and reverent mood, students will stand in awe of the revelation of the physical laws made visible in a new way.

TOPICS

The study of magnetism in the sixth grade treats three major topics:

- The attractive and repellent properties of magnets
- The polar orientation of magnets
- How magnetism travels

CHARACTERISTICS

In working with these topics, the following characteristics can be revealed:

- Magnets attract and are attracted to other magnets and to objects made of iron or steel.
- The attractive power of a magnet is concentrated at the poles.
- Unlike magnetic poles attract each other; like poles repel.
- Magnets have north- and south-seeking poles.
- Magnetism can travel through many materials.
- A field of force penetrates and surrounds a magnet.

If it is possible to have a five-week physics main lesson block in sixth grade, then a full week can be devoted to the study of magnetism and another week to the study of electricity. If only four weeks are available, teachers will need to limit the topics they cover in magnetism and static electricity.

The characteristics listed above can be revealed by some of the following demonstrations and activities.

DAY	DEMONSTRATIONS AND ACTIVITIES	CHARACTERISTIC
1	■ Magnetic Magnetite ■ What Types of Metals do Magnets Attract? ■ The Powerful Poles	■ Introduction to the attractive power of magnets. ■ Magnets attract and are attracted to other magnets and to objects made of iron or steel. ■ The attractive power of a magnet is concentrated at the poles.
2	■ Finding the North ■ Making Compasses	■ Magnets have north- and south-seeking poles.
3	■ Like and Unlike ■ Push and Pull ■ A Magnetic Spring	■ Unlike magnetic poles attract each other; like poles repel.
4	■ Magnetism Travels Through Solids ■ Without Wetting Your Hands ■ No Visible Means of Support	■ Magnetism can travel through many materials.
5	■ Lines of Force ■ Revealing Attraction and Repulsion	■ A field of force penetrates and surrounds a magnet.

Additional or alternative activities are included at the end of this chapter.

DAY 1

INTRODUCTION: THE ATTRACTIVE POWER OF MAGNETS

We can introduce the study of magnetism with a story of its discovery in ancient Greece or China. Such a story leads naturally to the demonstrations with natural magnets.

DEMONSTRATION

1. Magnetic Magnetite

CHARACTERISTIC: Magnets attract and are attracted to other magnets and to objects made of iron or steel.

A piece of magnetite or lodestone looks much like any other dark rock, but it has magnetic qualities.

MATERIALS

- 1 piece of magnetite
- other dark rocks
- small, light steel objects such as pins, staples, etc.
- 1 strong magnet
- 1 cardboard box

PROCEDURE

Test each of the pieces of rock to see if they attract the steel objects.

Put the magnetite in a box with the other rocks and use the strong magnet to locate it.

ACTIVITY

2. What Types of Metals Do Magnets Attract?

CHARACTERISTIC: Magnets attract and are attracted to other magnets and to objects made of iron or steel.

Through this activity, students will discover that a magnet only attracts other magnets and objects made of iron or steel.

PROCEDURE

Have students gather and test any materials that they think might be attracted by the magnet.

MATERIALS

1 magnet

an assortment of natural and man-made objects, including objects made of various metals

some smaller magnets

DEMONSTRATION

3. The Powerful Poles

CHARACTERISTIC: The attractive power of a magnet is concentrated at the poles.

Having seen that magnets attract iron and steel objects and other magnets, students can now determine where a magnet's power is concentrated.

PROCEDURE

Lower the magnet into a pile of the brads, tacks, or pins. Lift the magnet and observe whether the ends or the middle of the magnet attract the objects.

MATERIALS

1 bar magnet

several brads, tacks, or pins

COMMENT

During the review, introduce the word "pole," and explain that this is the part of the magnet where the magnetic power is concentrated.

DAY 2

DEMONSTRATION

4. Finding the North

CHARACTERISTIC: Magnets have north- and south-seeking poles.

We call the poles of magnets "north" and "south" because they point in those directions. After students have learned the laws of attraction and repulsion, they may ask, "How come the north pole of the magnet is attracted to the north pole of the Earth? Don't like poles repel each other?" We can explain that the poles of the magnet are named after the poles they seek, but for convenience we simply call them the north and south poles.

This demonstration could be set up before school so that when the students enter the classroom, they see several hanging magnets all facing the same direction. The orientation of the magnets can easily be disrupted, and it takes a while for them to come back to their position, so students should not touch the hanging magnets.

MATERIALS

several bar magnets
large paper clips
pieces of string, each about 2 feet long
1 map compass

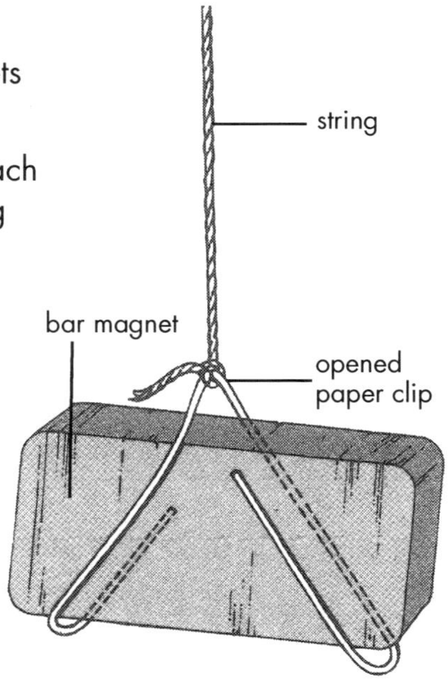

PROCEDURE

Make the paper clips into stirrups to hold the magnets. Attach a paper clip to the end of each string and attach the strings to the ceiling. Then mount a magnet on each stirrup. You now have magnets hanging from the ceiling. Make sure that the hanging magnets are at least a yard away from any objects made of iron. The magnets will begin to rotate and will eventually orient themselves in a north/south direction. Have the students check the magnets' orientation with a map compass.

ACTIVITY

5. Making Compasses

CHARACTERISTIC: Magnets have north- and south-seeking poles.

As a final activity for the second day, groups of students can make different kinds of compasses. Although this activity requires them to magnetize a piece of steel, the principles of magnetic induction will be dealt with later in the week.

PROCEDURE

To make a hanging compass: Magnetize the needle by repeatedly stroking the needle over one of the poles of the strong magnet, making sure always to move the needle in the same direction. Twenty to thirty passes should magnetize the needle adequately.

Tie a thread around the magnetized needle and adjust the knot so that the needle is balanced when it hangs. Drop a little melted wax or a little glue on the knot so that it will not slip. Tie the thread to a pencil that will rest over the mouth of the jar. Then let the thread hang down into the jar. Adjust the needle so that it does not touch the sides. The needle should point north and south. Check the orientation of the compass you have made with a map compass.

To make floating compasses: Magnetize the needle as described above. Place the magnetized needle in a piece of straw slightly longer than the needle and plug the ends of the straw with little balls of beeswax. Fill a bowl with water and float the straw in the bowl. The straw has now become a floating compass. Check its orientation with a map compass. The needle may need to be nudged away from the side of the bowl.

MATERIALS

- 1 strong magnet
- large sewing or darning needles
- candles
- liquid glue
- matches
- thread
- glass jars
- pencils
- plastic straws
- small pieces of beeswax
- medium-sized bowls
- water
- 1 map compass

DAY 3

DEMONSTRATION

6. Like and Unlike

CHARACTERISTIC: Unlike magnetic poles attract each other; like poles repel.

Although we might be tempted to let the students experience the push and pull of magnets as their first experience of attraction and repulsion, they can get so involved in that activity that the principles might not register fully. I have found it preferable to let the students first make visual observations of how the poles attract and repel each other and then experience the attraction and repulsion kinesthetically.

MATERIALS
2 bar magnets
paper clips
string

PROCEDURE
Make a stirrup for one of the bar magnets and hang it where it can swing freely. When it has stopped rotating, bring one end of the second magnet near to one pole of the hanging magnet and see whether that pole moves towards the second magnet or away from it.

Try the same procedure with the other pole. Students will clearly see how a pole of the hanging magnet is attracted to one pole of the other magnet and repelled by the other pole.

COMMENT
This can also be demonstrated by laying both bar magnets on a table. Two of the poles will quickly move towards each other, whereas two of the poles will push each other away.

ACTIVITY

7. Push and Pull

CHARACTERISTIC: Unlike magnetic poles attract each other; like poles repel.

MATERIALS

2 large strong magnets of similar size

PROCEDURE

Two students each hold one of the magnets and move them towards the other. (***CAUTION:*** *Make sure students' fingers are kept well away from the front surface of the magnet.*) As the magnets approach each other, students will feel either a strong force attracting or repelling the other magnet.

Have students try to hold the magnets close to each other but resist the power of the magnets. They will experience how much strength is needed to counter the magnets' push or pull.

ACTIVITY

8. A Magnetic Spring

CHARACTERISTIC: Unlike magnetic poles attract each other; like poles repel.

This activity can be done directly following the previous one because each student can then experience the phenomenon for him or herself. Alternatively, it can be done during the review the next day.

MATERIALS

4–5 disc magnets, each with a central hole

1 pencil with a diameter slightly smaller than the magnet's hole

PROCEDURE

Slide one of the magnets onto the pencil. Slide another magnet onto the pencil. If it is attracted to the magnet below, invert it so that like poles are facing each other and the upper magnet floats above the lower one. A series of repelling poles can be built up, and the group of magnets can act as a spring, compressing and expanding as the pencil is moved up and down.

COMMENT

Build a stand with a dowel on a base and put the magnetic spring on it. The spring can be kept in the classroom for a few days so that the students can play with it.

DAY 4

The following activities can be included if a full week is devoted to magnetism. If not, they can be done during the reviews later in the week, assigned as independent projects, or omitted.

DEMONSTRATION

9. Magnetism Travels Through Solids

CHARACTERISTIC: Magnetism can travel through many materials.

Like the other physical forces that have been studied, magnetism can travel through various materials including wood, glass, plastic, metal, and even through the human body.

MATERIALS

- pieces of wood, glass, and plastic
- 1 strong magnet
- 1 small steel object, preferably a steel ball

PROCEDURE

Place the steel object on one of the materials and hold the magnet beneath it. Move the magnet to see if it attracts the steel object and causes it to move. Test the other materials.

COMMENT

A steel ball or a spherical magnet is especially effective for this demonstration, for once it is attracted by the magnet, it will roll over the surface of the material but not roll off.

ACTIVITY

10. Without Wetting Your Hands

CHARACTERISTIC: Magnetism can travel through many materials.

This is a fun challenge: Can you remove a small steel object submerged in a cup of water without putting anything into the water or wetting your hands?

PROCEDURE

Hold the magnet to the side of or underneath the cup until the paper clip is attracted. Draw the paper clip out of the water by moving the magnet up the side of the cup. *(**CAUTION:** If students are using a glass, be sure to use a very light object. If the object is too heavy, the glass may break by the force of the object as it hits the side of the glass when it is attracted by the magnet.)*

MATERIALS

1 magnet

1 paper clip or light steel object

1 clear cup or glass

water

DEMONSTRATION

11. No Visible Means of Support

CHARACTERISTIC: Magnetism can travel through many materials.

This is one of the most delightful activities in the physics block, and students will want to continue playing with a "floating" paper clip because seeing an object suspended in the air with no visible means of support is thrilling.

MATERIALS

1 laboratory stand
1 ring clamp
1 strong magnet
1 paper clip
1 thread
masking tape

PROCEDURE

Place the magnet on the top of the ring clamp. It will be attracted to the clamp and stick to it without needing to be attached. Tie the thread onto one end of the paper clip and tape the other end of the thread to the table near the laboratory stand. Extend the thread with the paper clip toward the ring stand, then raise the clamp so that the magnet is positioned about ½ inch above the paper clip. The magnet will attract the paper clip and keep it suspended in mid-air. See how high the ring clamp can be moved up before the paper clip falls. Rotate the magnet a bit in various directions and the paper clip will follow it.

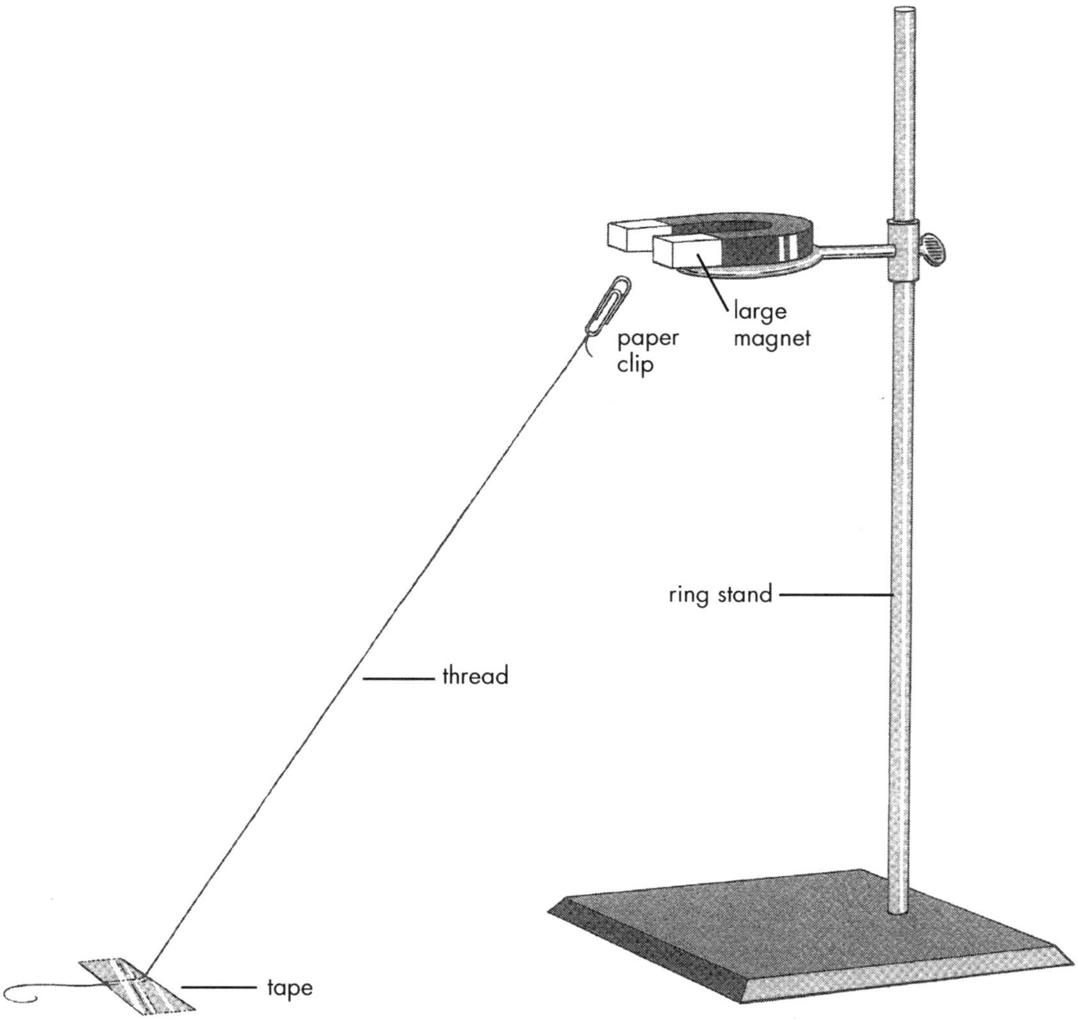

DAY 5

If a full week is being devoted to magnetism, the lines of force can be shown on the final day. If not, they can be demonstrated during the final review of magnetism.

DEMONSTRATION

12. Lines of Force

CHARACTERISTIC: A field of force penetrates and surrounds a magnet.

It is always good to end a week of physics with a strong and beautiful image. Like the Chladni plate, the colored shadows, and the convection currents in water, illustrating the lines of magnetic force will remain imprinted in the students' minds.

Magnetism radiates outward from a magnet in a three-dimensional field of force. This field can be made visible by iron filings. When demonstrating the field of force, work slowly and allow the phenomenon to be revealed gradually so that students can fully experience the wonder of making the invisible visible.

PROCEDURE

Lay the magnet flat on a surface, place the piece of cardboard over it, and cover the cardboard with the piece of white paper. Sprinkle the iron filings onto the paper. The lines of force will become visible.

COMMENT

This demonstration can also be done with magnets of other shapes, but too many examples can diminish the power of this experience.

MATERIALS

1 bar magnet
1 piece of cardboard
1 pane of Plexiglas or glass
1 piece of white paper
iron filings in a salt shaker

DEMONSTRATION

13. Revealing Attraction and Repulsion

CHARACTERISTIC: A field of force penetrates and surrounds a magnet.

As with the previous demonstration, do this one slowly so that the magnetic field is gradually revealed.

MATERIALS

- 2 strong bar magnets
- 1 piece of cardboard or a pane of Plexiglas or glass
- 1 piece of white paper
- iron filings in a salt shaker

PROCEDURE

Lay the magnets flat on a surface, parallel to each other so that the north end of one is aligned with the south end of the other, about an inch apart. Cover the cardboard, Plexiglas, or glass with a piece of paper and put it on top of the magnet. Sprinkle the iron filings on the paper. The lines of force of the two magnets and the attractive force between them will be visible.

To show the force of repulsion, place the magnets so that the north poles and south poles are aligned with each other. Repeat the procedure described above.

COMMENT

If the magnets attract each other too strongly and do not stay apart, you can separate them by a small piece of wood or plastic. Alternatively, they can be taped in place.

ADDITIONAL OR ALTERNATIVE ACTIVITIES

ACTIVITY

14. Polar Power

CHARACTERISTIC: The attractive power of a magnet is concentrated at the poles.

In this activity students use paper clips to test which part of the magnet can hold the most weight. If time does not allow this to be done in class, some students could do it as an independent project and then share the results with the class.

PROCEDURE

Bend the large paper clip into a "J" shaped hook and attach its straight stem to the pole of a bar magnet. Hold the magnet in the air and hang as many small paper clips from the curved end of the hook as the magnet can hold. When the hook falls, count how many paper clips the magnet could hold.

Repeat this procedure attaching the long stem of the hook to the other pole of the magnet. Compare how many clips each pole can hold.

Attach the hook to the center of the magnet and test to see how many clips can be held.

Repeat this activity with a horseshoe magnet.

MATERIALS

1 bar magnet
1 horseshoe magnet
1 large paper clip
several small paper clips

INVESTIGATION

15. Can Magnetism Be Screened?

CHARACTERISTIC: Magnetism can travel through many materials.

MATERIALS

1 strong magnet
1 paper clip
1 piece of string
masking tape
1 laboratory stand
1 ring clamp
thin sheets of various materials such as paper, wood, glass, copper, or steel

Different substances screen magnetism to a greater or lesser extent. Substances that are non-magnetic allow magnetism to pass through them unimpeded, while those like iron or steel screen magnetism quite effectively.

PROCEDURE

Set up the stand, clamp, and magnet as in *Demonstration 11: No Visible Means of Support*. Try putting the sheets of different materials between the magnet and the paper clip to see which of them interfere with the magnetic attraction.

ACTIVITY

16. The Three-Dimensional Field of Force

CHARACTERISTIC: A field of force penetrates and surrounds a magnet.

MATERIALS

1 small jar with a tight-fitting lid
iron filings
light salad oil or mineral oil
2 strong magnets

PROCEDURE

Fill the jar completely with oil, pour in some iron filings, and cap the jar. Shake the jar so that the iron filings are well mixed in the oil. Set the jar on a flat surface and hold one pole of a magnet against one side of the jar. The filings will display three-dimensional lines of the field of force.

Repeat this procedure holding the poles of two magnets on opposite sides of the jar. The patterns of the field will vary depending on whether you use like or unlike poles of the magnets.

The Study of Electricity in the Sixth Grade

We conclude our introduction to physics in the sixth grade with the study of electricity. When we studied magnetism, we felt that we were entering the earthly realm of minerals and metals; with electricity we approach more mysterious realms.

We have seen how the sense organs by which we perceive sound, light, heat, and magnetism work from the inside of our bodies toward the surface, giving us an increasingly physical experience of the forces. We have no specific sense organ designed to perceive electricity; rather, electricity stimulates our sense organs directly and can affect our senses of touch, taste, smell, seeing, and hearing.

Because the phenomena of static electricity are more accessible and comprehensible than those of current electricity, we introduce static electricity in the sixth grade and move on to current electricity in the seventh. The distinctions between static and current electricity are complex and difficult to understand. For instance, the electricity generated by rubbing is not static at all; it is the same force as the one we will create in seventh grade by chemical means or by moving a coil within a magnetic field. Though electricity generated by rubbing is technically called triboelectricity—from the Greek root *tribein*, which means "to rub"—it is just another aspect of electromagnetism.

The causes of static electricity are not easily understood, but sixth graders don't need to understand them; they just need to experience how static electricity can be generated.

In the classroom, we might begin the study of static electricity by describing its discovery by the Greek scientist Thales. He found that when he rubbed amber with a woolen cloth, the amber developed the power to attract small light objects. We might also point out that the word electricity comes from the ancient Greek word *electron*, which means "amber." Although amber can develop a weak electric charge, there are modern substances that develop a far stronger charge. In our study of magnetism, we saw how the force is induced in earthly substances—minerals and metals. Static electricity is most strongly induced in man-made, synthetic substances such as polyethylene, acrylic, and acetate.

Once we can generate static electricity, we can begin to study its effects. These demonstrations and activities will lead students to conclude that static electricity and magnetism have much in common: both forces can be generated by rubbing objects; both have the power of attracting objects; and both express themselves in polarities of attraction and repulsion.

On a deeper level, magnetism and electricity have even more in common, and some of these connections will become apparent in the study of electromagnetism in the seventh and eighth grades.

TOPICS

The study of electricity in the sixth grade treats three major topics:

- The generation of static electricity
- The detection of static electricity
- Properties of static electricity

CHARACTERISTICS

In working with these topics, the following characteristics will be revealed:

- Static electricity is produced by rubbing certain substances with other substances.
- A static electric charge has the power of attraction or repulsion.
- Static electric charges can be positive or negative.
- Like charges repel each other and unlike charges attract each other.
- Static electricity can be detected and stored.

If a full week is allowed for the study of electricity, the topics and characteristics listed above can be introduced by the following demonstrations, activities, and investigations. If magnetism and electricity are covered in the same week, you may only have two days to cover electricity, so you will have to select which topics should be omitted.

DAY	DEMONSTRATIONS AND ACTIVITIES	CHARACTERISTIC
1	■ Amber's Powers of Attraction ■ Generating Static Electricity	■ Static electricity is produced by rubbing certain substances with other substances.
2	■ The Attractive Power of Static Electricity	■ A static electric charge has the power of attraction or repulsion.
3	■ Attraction and Repulsion ■ Repulsion Between Like Charges ■ Reverse the Charges	■ Static electric charges can be positive or negative. ■ Like charges repel each other and unlike charges attract each other.
4 & 5	■ Making an Electroscope ■ What Kinds of Objects Can Develop an Electrostatic Charge? ■ Using an Electrophorus ■ Making a Leyden Jar	■ Static electricity can be detected and stored.

Additional or alternative activities are included at the end of this chapter.

DAY 1

DEMONSTRATION

1. Amber's Properties of Attraction

CHARACTERISTIC: Static electricity is produced by rubbing certain substances with other substances.

Static electricity might be introduced by telling the story of its discovery by Thales, the ancient Greek scientist, who discovered that when he rubbed amber with a woolen cloth it attracted small, light objects.

If it isn't possible to find any amber, move on to the next demonstration.

MATERIALS

1 piece of amber

1 woolen cloth

several small pieces of paper, wood shavings, small feathers, or threads

PROCEDURE

Rub the amber with the piece of woolen cloth. The amber will attract small objects when held near them.

DEMONSTRATION

2. Generating Static Electricity

CHARACTERISTIC: Static electricity is produced by rubbing certain substances with other substances.

Static electricity can be generated by rubbing polyethylene, vinyl, or rubber with woolen cloth or fur; or by rubbing acrylic, Plexiglas, or glass with silk. The synthetic materials will develop a charge even when there is a good deal of moisture in the air.

PROCEDURE

Charge the rod by rubbing it vigorously with the woolen cloth or fur. Dangle the strip of paper from one hand and bring the charged rod near it. The paper will be attracted to the rod.

Sprinkle Styrofoam pellets on a table top. Charge the rod again and pass the rod over the pellets; they will be attracted to it.

Use the charged rod to attract a piece of string or a balloon. The rod can also be brought near a person's hair, which will rise to meet it.

MATERIALS

- 1 rod or strip of polyethylene
- 1 piece of woolen cloth or fur
- 1 strip of paper
- Styrofoam packing pellets
- 1 piece of string
- 1 inflated balloon

DEMONSTRATION

3. Generating Static Electricity, Continued

CHARACTERISTIC: Static electricity is produced by rubbing certain substances with other substances.

PROCEDURE

Repeat the procedures from the previous demonstration and see whether this type of rod is also able to attract the objects.

MATERIALS

- 1 rod or strip of acetate or glass
- 1 piece of silk cloth
- 1 strip of paper
- Styrofoam packing pellets
- 1 piece of string
- 1 balloon

DAY 2

DEMONSTRATION

4. The Attractive Power of Static Electricity #1

CHARACTERISTIC: A static electric charge has the power of attraction or repulsion.

Attraction caused by static electricity can be demonstrated in countless different ways. Be selective about which demonstrations to do so that students don't become overwhelmed.

MATERIALS
1 inflated balloon
1 piece of woolen cloth

PROCEDURE

Rub the balloon vigorously with the wool. When put against a wall or a surface, the balloon will stick to it. When put against a person's clothes, it will stick to most fabrics.

DEMONSTRATION

5. The Attractive Power of Static Electricity #2

CHARACTERISTIC: A static electric charge has the power of attraction or repulsion.

MATERIALS
several strips of newspaper
1 piece of soft cloth

PROCEDURE

Hold a strip of newspaper against a smooth vertical surface such as a blackboard. Rub the newspaper with the cloth. It will stick to the board.

DEMONSTRATION

6. The Attractive Power of Static Electricity #3

CHARACTERISTIC: A static electric charge has the power of attraction or repulsion.

PROCEDURE

Tie the string to the inflated balloon and hang the balloon from the ceiling. Charge the rod and bring it near the balloon. It will be attracted to the rod.

COMMENT

This demonstration can also be done by bringing a charged rod near a soap bubble that is floating in the air.

MATERIALS

- 1 small balloon of the sort used to make water balloons
- 1 piece of string
- 1 acetate or glass rod
- 1 piece of silk
- soap for blowing soap bubbles

DEMONSTRATION

7. The Attractive Power of Static Electricity #4

CHARACTERISTIC: A static electric charge has the power of attraction or repulsion.

A charged rod can also deflect a stream of water. This is an unexpected and impressive phenomenon.

PROCEDURE

Punch a small hole in the side of the plastic bottle near the bottom and cover the hole with masking tape. Fill the bottle with water and elevate it about 12 inches above the side of the tray. When the tape is removed, the water will flow in an arc onto the tray. When the charged rod is brought near the water, the water stream will be attracted to the rod.

MATERIALS

- 1 polyethylene or acetate rod
- wool or silk cloth
- a 1- or 2-liter plastic bottle
- 1 small hole punch or drill
- masking tape
- 1 rimmed tray or pan
- water

COMMENT

If a faucet is handy, this can also be done with a fine stream of water running from the faucet.

DEMONSTRATION

8. The Attractive Power of Static Electricity #5

CHARACTERISTIC: A static electric charge has the power of attraction or repulsion.

MATERIALS

- 1 piece of Plexiglas or glass
- 1 piece of silk cloth
- several thick books or pieces of wood
- puffed rice cereal or little pieces of Styrofoam

PROCEDURE

Make two stacks of books or wood with a space between them so that they hold the Plexiglas or glass suspended over the table. Sprinkle the puffed rice or pieces of Styrofoam underneath the glass. As you rub the top of the Plexiglas with silk, the objects below will be attracted to it.

COMMENT

After the little objects fly up to the Plexiglas, they remain attached for a while and then are suddenly repelled. This principle will be explored further when attraction and repulsion are studied more specifically.

INVESTIGATION

9. What Kinds of Materials Will Be Attracted to a Charged Object?

CHARACTERISTIC: A static electric charge has the power of attraction or repulsion.

This investigation can be done as an independent project and the results shared with the class.

PROCEDURE

Hang some of the light materials by threads from the ring clamps. Make sure the materials can swing freely. Charge the polyethylene rod by rubbing it with woolen cloth or fur and bring it near one of the hanging materials to see if it is attracted, is repelled, or makes no movement. Repeat with the lightweight objects set on a table and record the results. Charge the acetate or glass rod and test the different materials and objects. Compare the results from the two rods.

MATERIALS

several laboratory stands and ring clamps
1 polyethylene rod
1 piece of woolen cloth or fur
1 acetate or glass rod
1 piece of silk or acetate fabric
a variety of thin, light materials that can be hung, such as paper, thread, wire, plastic, aluminum foil, Mylar tape, etc.
a variety of lightweight objects such as paper, Styrofoam, seeds, dust balls, etc.
sewing thread

DAY 3

Objects that carry an electrostatic charge behave much like magnets with charged north and south poles. Like charges repel and unlike charges attract. In some of the previous activities, it was shown that a charged object can attract an uncharged one. The following demonstrations show how objects with like charges are repelled.

DEMONSTRATION

10. Attraction and Repulsion

CHARACTERISTIC: Like charges repel each other and unlike charges attract each other.

This demonstration allows students to deduce that the different materials develop different charges, that objects with like charges are repelled, and that those with unlike charges are attracted to each other.

PROCEDURE

Drill a hole in the center of the Plexiglas strip so that it will balance when hung. If it doesn't balance, weight the lighter end with some tape. Hang the Plexiglas strip from the ceiling or from a laboratory stand. Charge the Plexiglas piece with the woolen cloth or fur. Charge

MATERIALS

1 polyethylene rod
1 piece of woolen cloth or fur
1 glass or acrylic rod
1 piece of silk or acetate fabric
1 rectangular strip of Plexiglas about 1″ x 6″
1 drill with a small bit
string
1 laboratory stand
1 ring clamp

the polyethylene rod with the woolen cloth or fur. Charge the glass or acrylic rod with silk or acetate. Bring one of the charged rods near the Plexiglas and see if it swings toward the charged rod or away from it. Bring the other charged rod near the Plexiglas and note the results.

DEMONSTRATION

11. Repulsion Between Like Charges

CHARACTERISTIC Like charges repel each other and unlike charges attract each other.

Water balloons filled with air are just the right size and weight for activities with static electricity.

MATERIALS

- 2 small lightweight balloons, such as those used for water balloons
- string
- masking tape
- 1 piece of woolen cloth or fur

PROCEDURE

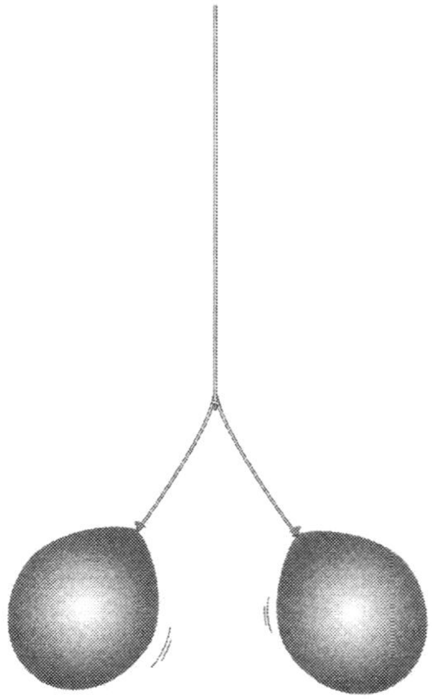

Inflate the balloons and tie them to each end of a piece of string. Tie the string to another strand so that they form an upside down Y, with the two arms of the Y being about 12 inches long. Attach the stem of the Y to the ceiling with masking tape.

Charge one of the balloons by rubbing it with woolen cloth or fur and then let go of it. Observe what happens when it comes near the other balloon. Now charge both balloons. When they come near each other, they will repel each other and will not make contact for a long time.

DEMONSTRATION

12. Reverse the Charges

CHARACTERISTIC: Like charges repel each other and unlike charges attract each other.

An uncharged object can receive a charge from a charged object. Once it has received that charge, the two similarly charged objects repel each other.

MATERIALS

- 1 polyethylene rod
- 1 piece of woolen cloth or fur
- 1 glass or acrylic rod
- 1 piece of silk or acetate fabric
- 1 pith ball or small bead of Styrofoam hanging from a thread
- 1 laboratory stand
- 1 ring clamp

PROCEDURE

Charge the polyethylene by rubbing it with woolen cloth or fur. Bring the rod near the ball. The ball will be attracted to the rod. After a few moments, the ball will fly off the rod. If the rod is brought close to the ball again, it will be repelled by the rod.

Neutralize the charge on the ball by holding it between your fingers. Repeat the procedures, this time with the glass or acrylic rod charged with the silk or acetate fabric. The ball will again be attracted to the rod and then be repelled.

Now charge the two rods with the appropriate materials and hold them on either side of the ball. The ball will be attracted to one of the rods. When the ball is repelled from that rod, it is attracted to the other rod. When it is repelled again, it will be attracted to the first rod.

DAYS 4 & 5

ACTIVITY

13. Making an Electroscope

MATERIALS
- 1 polyethylene rod
- 1 piece of woolen cloth or fur
- 1 glass or acrylic rod
- 1 piece of silk or acetate fabric
- 1 piece of thread
- 1 needle
- 1 bead of Styrofoam or piece of puffed rice
- 1 laboratory stand or a wire hanger bent into the shape of a stand

CHARACTERISTIC: Static electricity can be detected and stored.

An electroscope is an instrument that indicates the presence of an electrostatic charge. The simple electroscopes that do not indicate the polarity of the charge are most appropriate for sixth grade.

PROCEDURE

Thread the needle and tie a knot at the end of the thread. Push the needle through the Styrofoam bead or puffed rice and pull it through until the knot is flush against the bottom of the bead. When a charged rod or other charged object is brought near the charge detector (electroscope), the piece of foam or rice will be attracted to it.

INVESTIGATION

14. What Kinds of Objects Can Develop an Electrostatic Charge?

CHARACTERISTIC: Static electricity can be detected and stored.

Use the electroscope made in the previous activity or purchase one from a science supply company. This investigation can be done as an independent project and its results reported to the class.

PROCEDURE

Test the electroscope by using a substance that you know can be charged with static electricity. Test each of the other substances by rubbing it with fur or wool and putting it near the electroscope. Record the results. Then rub each object with the silk or acetate and repeat the tests. Record the results.

MATERIALS

- a variety of objects such as wood, metal, plastic, glass, rubber
- 1 piece of fur or wool
- 1 piece of silk or acetate fabric
- 1 electroscope

DEMONSTRATION

15. Using an Electrophorus

CHARACTERISTIC: Static electricity can be detected and stored.

The electrophorus—an instrument invented by Alessandro Volta—is capable of producing a number of successive electrostatic charges. It consists of a charging plate and a charge transfer disk attached to an insulated handle. An electrophorus can be bought from a scientific supply company, but a simple one, can be made as follows:

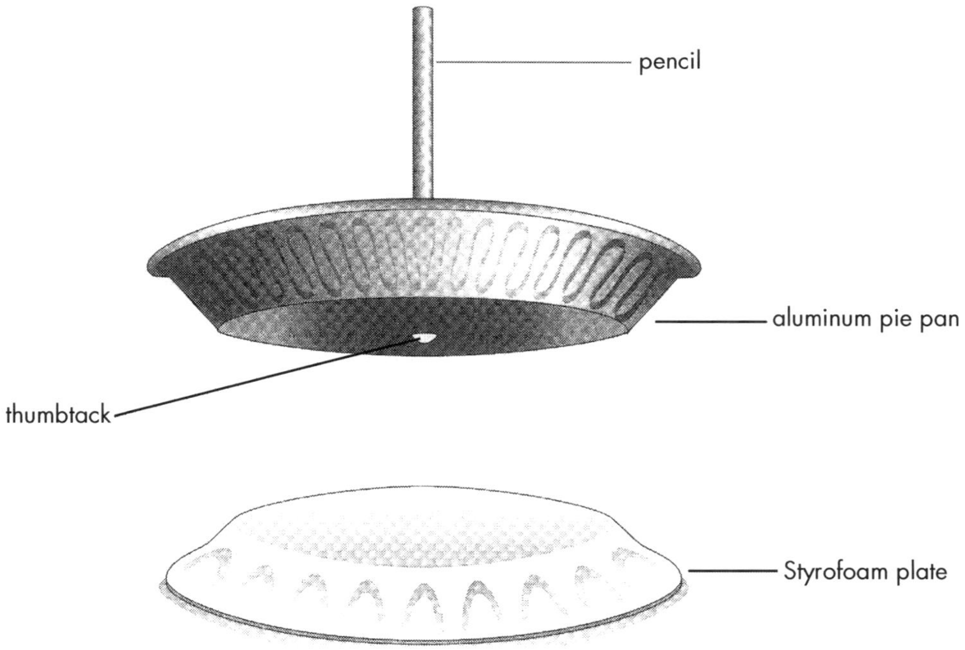

MATERIALS

1 aluminum pie pan
1 pencil with a rubber eraser
1 thumb tack
1 Styrofoam plate
1 woolen cloth

PROCEDURE

Make a handle for the pie pan by pushing the thumb tack through the pan and into the pencil's eraser. Place the Styrofoam plate upside down on a surface and rub it vigorously with the woolen cloth for a minute. Hold the pie pan by the handle and push it firmly against the bottom of the upside down Styrofoam plate. While the pie pan is still resting on the Styrofoam, touch the edge of the pie plate with one finger. When you lift the pie pan off the plate, it will be highly charged.

To discharge the electrophorus, bring your knuckles near the pie pan and you will feel, see, and hear a small spark. Once the electrophorus has been discharged, it can be charged again by simply putting it back on the plate and touching it with your finger. The plate does not need to be rubbed to recharge it each time.

COMMENT

An old-fashioned vinyl LP record can be used instead of a Styrofoam plate. A Wimshurst machine is essentially two rotating electrophoruses that produce a powerful succession of sparks. It is available from scientific supply companies.

ACTIVITY

16. Making a Leyden Jar

CHARACTERISTIC: Static electricity can be detected and stored.

A Leyden jar is a device that can store fairly large electrical charges. A Leyden jar may be purchased from a science supply company, but this simple one, will serve most purposes

PROCEDURE

Wrap the aluminum foil around the bottom and the outside of the pill bottle, going about two-thirds up the side. If necessary, tape the foil in place. Fill the pill bottle almost full of water and cap it. Insert a nail through the cap and push it down so that it touches the water and its head protrudes about ¾ inch above the cap.

To charge the Leyden jar, charge an electrophorus or a polyethylene rod and bring it near the nail. A small charge will jump to the nail. To discharge the Leyden jar, hold the outer foil and touch the nail. Depending on the number of charges, a shock can be felt.

MATERIALS

- 1 small plastic pill bottle with a pop-off cap
- 2-inch finishing nail
- aluminum foil
- adhesive tape
- water

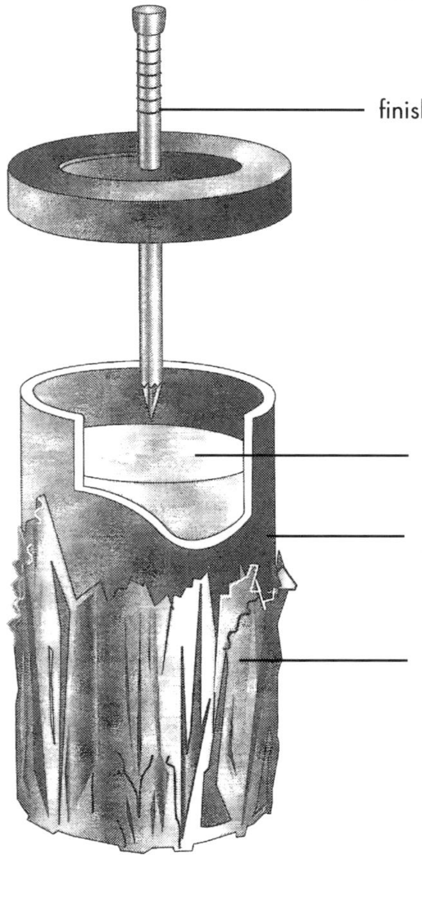

Notes

Teacher Resources

WALDORF EDUCATION AUTHORS

Baravalle, Hermann von. *Introduction to Physics in the Waldorf Schools*. Englewood, N.J.: Waldorf School Monographs, 1967.

Bojarsky, Mikko & Antje. *Mechanics: A Demonstration Manual for Use in the Waldorf Seventh Grade Physics Lesson.* Mikko Bojarsky, 2009.

Bojarsky, Mikko. *A Demonstration Manual for Use in the Waldorf School Eighth Grade Physics Main Lesson.* Mikko Bojarsky, 2009.

D'Aleo, Michael & Edelglass, Stephen. *Sensible Physics Teahing*. New York: Parker Courtney Press, 1999.

Fairman, Eric. *A Path of Discovery. Grades 6, 7, 8.* Sidney, Australia: Eric Fairman, 2001, 2002.

Julius, Fritz. *The World of Matter and the Education of Man.* Forest Row, Sussex: Steiner Schools Fellowship Publications.

Lehrs, Ernst. *Man or Matter.* London, England: Rudolf Steiner Press, 1951.

Mackensen, Manfred von. *A Phenomena Based Physics, Vols. 1–3.* AWSNA Publications.

Proskauer, Heinrich O. *The Rediscovery of Color.* Spring Valley, NY: Press, 1986.

Sonner, Robert, H. *A Compendium of Phenomenological Experiments: Teachers Reference Book*. Mission Hills, CA: Robert H. Sonner, 2013.

Stockmeyer, E. A. Karl. *Rudolf Steiner's Curriculum for Waldorf Schools.* UK: Steiner Waldorf Schools Fellowship, 2001.

Wilkinson, Roy. *The Physical Sciences I.* Forest Row, Sussex: Roy Wilkinson, 1977.

RUDOLF STEINER REFERENCES

Steiner, Rudolf. *A Modern Art of Education. Lecture l0.* Great Barrington, MA: Anthroposophic Press, 2004.

——. *Practical Advice to Teachers. Lecture 8*. Great Barrington, MA: Anthroposophic Press, 2000.

——. *Soul Economy and Waldorf Education. Lecture 11.* Great Barrington, MA: Anthroposophic Press, 2003.
——. *The Kingdom of Childhood. Lectures 3 and 7.* Hudson, NY: Anthroposophic Press, 1995.
——. *The Light Course.* Great Barrington, MA: Anthroposophic Press, 2001.
——. *The Renewal of Education. Lectures 8 and l0.* Great Barrington, MA: Anthroposophic Press, 2001.
——. *The Spiritual Ground of Education. Lecture* 6. Great Barrington, MA: Anthroposophic Press, 2004.
——. *The Warmth Course.* Spring Valley, NY: Mercury Press, 1988.
——. *Education for Adolescents. Lectures 1–4.* Hudson, NY: Anthroposophic Press, 1996.

SCIENCE SUPPLY COMPANIES

Carolina Biological Supply Company
2700 York Road, Burlington, NC 27215-3398
www.carolina.com

Flinn Scientific
P.O. Box 219, Batavia, IL60510
www.flinnsci.com

Sargent Welch
PO Box 92912
Rochester, NY 14692-9012
sargentwelch.com

17270398R00083

Printed in Great Britain
by Amazon